The Arguments
of Agriculture

Science and Society:
A Purdue University Press Series
in Science, Technology, and Human Values

Leon E. Trachtman, General Editor

Volume 7

The Arguments of Agriculture

A Casebook in Contemporary Agricultural Controversy

by Jan Wojcik

Purdue University Press
West Lafayette, Indiana
1989

Second printing February 1991

Library of Congress Cataloguing-in-Publication Data
Wojcik, Jan, 1944–
    The arguments of agriculture : a casebook in contemporary agricultural controversy / by Jan Wojcik.
        p.   cm. — (Science and society ; v. 7)
    Bibliography: p.
    Includes index.
    ISBN 0–911198–99–7 :
    1. Agriculture—United States.   2. Agricultural innovations—United States.   I. Title.   II. Series: Science and society (West Lafayette, Ind.) : v. 7.
    S441.W85   1989
    630′.973—dc19   88–30488

Printed in the United States of America

For Mauri Williamson

frigoribus parto agricolae plerumque fruuntur
mutuaque inter se laeti conviva curant

It is when cold winds flow that farmers share
the fruits of their harvests in joyful gatherings.

Virgil
Georgics I: 300-301

Contents

Preface

The Order of Arguments

It is an interesting time to be leaning over the fences of American farms. There are discussions, even arguments, in the land about whether farmers ought to change the way they farm. Should agriculture continue to industrialize? One can hear good reasons why it should. Fewer hands working with better tools on larger farms hold out the promise of lower prices for food and perhaps even greater variety. It has now become routine for supermarkets to offer June Bearer strawberries in January. Or alternatively, should agriculture slow down, diversify, break up into small holdings worked by the hands of their owners? One can hear good reasons why it should. More farmers working carefully with appropriate tools hold out the promise of preserving the wholesomeness of the food they grow and the land. Better a roadstand homegrown apple in October than a supermarket ethylene-ripened tomato in March.

There have been arguments like this heard before. As the discussion in chapter 4 explains, one hundred and fifty years ago American farmers debated among themselves whether they should adopt the progressive English methods of farming: spreading manure and rotating different crops and animals onto their fields. Before the 1820s farming in America was a way of life for most farmers. Although it barely provided a subsistence, farming offered a community for like-minded people who lived and often worshipped together. Between the 1820s and 1862, when Congress passed and Abraham Lincoln signed the Morrill Act establishing the land-grant system of agricultural universities, farming became more professional and more competitive. Growing city populations needed more food. Farmers needed to crop the same fields year after year, especially if they were close to cities or navigable rivers. The more progressive farmers decided they could no longer depend on the questionable luxury of clearing more land on the outskirts when their original farm lost its good tilth. Their arguments and their style of farming carried the day. We could say its legacy continues with the development of every new chemical, machine, or technique for raising crops and animals more efficiently on the same fields.

The new debate resembles the earlier one because it is fundamental. Should we again change the definition of farming? Once it was thought to be primarily a way of life. Then it became a profession that was still a way of life for most farmers who continued to live on the land they worked. If it now becomes an industry, will it become only a profession? Is it sentimental or sensible to think it must remain a way of life if it is to remain true to its nature and true to Nature herself?

The new debate differs from the earlier one by being more vociferous, if not more raucous. There are more voices in it, and not all of them are the voices of farmers. Many of them, in fact, belong to the children or grandchildren of the decreasing number of increasingly productive farmers, each one of whom grows food for more than fifty fellow citizens off the farm. Those who no longer need to farm can make their living as scientists, engineers, industrial workers, shopkeepers, politicians, or teachers. But many of them have passionate interests, even vested interests, in how or whether the farming of their fathers and mothers should change. This book presents some of their ideas.

It examines five arguments being used today to wrestle with the question, How should farmers farm? From time immemorial there has been really only one answer. Farmers should farm so that they can farm again. They need to produce more than healthy food for their families and fellow citizens; they need to cultivate and sustain the health of the soil that produces the food. The same fields must produce food dependably year in and year out. But today modern farm tools and chemicals take huge and unprecedented yields from those fields. They often exact a cost in an equally unprecedented loss of healthy topsoil and farmers from their farms. A growing international market for food and fiber puts mounting pressure on farmers all over the world, even in traditional cultures, to use ever more sophisticated techniques to push their productivity higher—and at the same time, to raise the risk of their losses. Under this kind of pressure, the basic question about farming splits into many smaller ones. The answers multiply and become contradictory. Hence this effort to sort the questions onto different shelves, the answers into different bins. It supplies a primer in the basic arguments of agriculture.

Each chapter describes a different argument. Chapter 1 discusses various general theories about the adversarial nature of agriculture. Chapter 2 examines the two basic principles behind every agricultural debate. Chapter 3 relates the basic principles to two fundamental policies according to which politicians strive to put the principles into practice. Chapter 4 examines the perils and promise of agricultural technology. Chapter 5 examines the benefits and dangers in the international trade of agricultural technologies and goods. The chapters are arranged in order of the increasing complexity of their topics.

The book sticks to the basics. There is little quantitative data or statistical fact beyond what is required to clarify an argument under review. Readers are referred to the readings listed at the end of each chapter for fuller discussions of these matters. Instead, the body of each chapter describes the various arguments about agriculture and features the fundamental attitudes behind them. The author believes that it is precisely these attitudes which determine what kind of statistical support a researcher will seek. Agriculture is a partisan science. Sincere experts, armed with wonderful statistics, can be brought to the defense of any perspective on what agriculture is or what it should be. They are all given an equal opportunity here to voice their opinions.

But the book is not intended to cultivate scholarly detachment. These arguments underlie real-life issues affecting the daily lives and livelihood of farmers—and all citizens—in the modern world. To make this clear and to invite readers to puzzle out well-informed personal opinions, each chapter begins with a section titled "The Cases." This section consists of three or more short narratives illustrating the kinds of theoretical issues discussed in the chapter. The stories are all made up, but true to life. Their details and facts derive from real stories to be found in newspaper accounts, academic studies, books, and popular magazine articles. The stories are true to life in another way as well. Each one concerns a conflict. Someone or some group is under pressure to make a specific decision about farming, without being sure what the outcome will be. This is true to life because the sum total of such decisions, made in private or made in public, determines how agriculture changes in this day or any other. That is why it is important for each one of us, whatever we have to do with farming, to understand the arguments of agriculture as best we can. They are the seeds of our future harvests, and as many thinking people as possible should have a hand in their sowing.

To help focus some thought along these lines, a section entitled "The Issues" follows "The Cases" in each chapter. It provides a broad overview of the kinds of arguments used in deciding real-life cases. This section introduces a more detailed section titled "The Review," which makes up the bulk of the chapter. The arrangement of this third section varies, but in each chapter the author takes care to present a wide range of differing opinion clearly and fairly. Where appropriate a chapter ends with a conclusion that summarizes and sometimes seeks a compromise among the various arguments in the chapter. The goal is always to help the reader understand the various points of the argument at hand. It is, of course, impossible for any one book to present every possible nuance of thought on any issue pertaining to a topic as large and as complicated as agriculture. The reader is offered here a selective sketch that suggests the wide range of opinion possible on any one issue.

The author, of course, has strong favorites among these many arguments. In general he favors careful agriculture, smaller farms, organic methods, and owner-operators. But many colleagues and writers with different opinions have made him thoughtful where, left to himself, he would be certain. The author hopes that each reader will find his or her own favorite arguments—whatever they are—honestly presented in these pages. At the same time, he hopes that they will come to share his respect for the best among the opposing arguments of agriculture. Perhaps the best preamble to discovering what one really thinks or really wants to do is to experience a well-informed ambiguity—at least for a moment. As an exercise in impartiality, each reader might try to imagine participating in each case. He or she could adopt the role of one of the characters in the drama, trying to fathom why that character says and acts as he or she does. After reading each case, ask the question, Was this right to do? Could another decision have been made? Perhaps one reader will decide nothing else could have been done; the outcome was inevitable. Another might wish the decision had gone another way. In every case, a reader will probably have strong opinions in those stories that come closest to home, describing events similar to those the reader has witnessed, read about with interest, or even participated in. The book works best if the reader comes to a provisional conclusion about what happened in the stories before reading further in the chapter for an account of what other people think about situations similar to those depicted in the stories. This way the reader can test his or her thinking against the best thinkers engaged in the public arguments of agriculture today. Even if a reader does not change his or her mind, perhaps he or she will gain a new appreciation for the honesty and validity of opposing points of view.

There is only one new idea developed here: there are really no new ideas in arguments about agriculture. As the discussion at the beginning of chapter 2 suggests, when the Roman Republic became an empire, her agriculture changed from a subsistence way of life for families on small seven-acre farms to a major industry powered by slaves (the big machines of the day) on *latifundia* (literally "big spreads"—the agribusiness farms of the day). Today, under the surface of often sophisticated, political arguments in favor of erosion control or loans to save the family farm or protective subsidies or even quotas and embargoes for foreign produce, we find at their roots the same basic arguments certain noble Romans used to caution their countrymen about abandoning the goddess of corn, Ceres, to the winds of free trade.

In sum, the reader makes best use of this book by using its cases and reviews to appreciate the best thinking behind every argument made about agriculture. We all have a vested interest in making these arguments as clear and precise as they can be. In debate, as in nature, the old Eskimo adage holds: "The wolf makes the caribou strong." This means the honest give-and-take of every good argument improves the understanding of each participant. Finally, each of us, on each side of every one of the arguments of agriculture, wants an answer to the basic question with which we begin: How should farmers farm to keep themselves, their land, and their fellow citizens in good health? No eater is ever far afield.

Classroom Use

If this book is used in a classroom, different students might be assigned to the roles of different characters in each case. They could act out and elaborate each episode, filling in dialogue as they might imagine each character would speak, while making use of the ideas and arguments reviewed in the rest of the chapter. After a certain point in the dialogue, other class members could join in and take sides. They might bring up points one of the participants missed. Or they might vote on which participant was more convincing and then discuss why. A faculty member could provide each participant with outside reading or appoint "expert advisors" to do special research and help a participant in preparing an argument.

Acknowledgments

A grant from the Kellogg Foundation supported the writing of this book. It was first used in manuscript as a text in a course entitled "Dilemma and Decision in Modern Agriculture," the development and the teaching of which the Kellogg Foundation also supported. James J. Vorst, professor of agronomy, and Richard Blanton, professor of sociology and anthropology at Purdue University, team taught the course. Both they and their students made many suggestions about how to make the manuscript more useful to them. Martha Bailey, librarian, made finding obscure writings easy. The generous kindness of them all is imprinted on every page.

Whatever is to be learned about agriculture in these pages is due to the patience of my teachers. First and foremost is my wife, Christine Zavgren, who grew up on a farm in Nebraska, where she learned the productive intimacies of landscape, animals, people, and weather. Charles Boebel, professor of English at Manchester College, was himself a child of farmers and shared with me over many a fence his enduring love of the land. Mauri Williamson, the only executive secretary the Purdue Agricultural Alumni Association has ever needed, convinced a lot of Indiana farmers they could teach an English professor in a tweed coat a lot about farming if they were patient. Rudy Hilst, the former associate dean of the School of Agriculture, encouraged faculty to cross the street that separates the School of Agriculture from the School of Humanities, Social Sciences, and Education at Purdue—in both directions. James Vorst, professor of agronomy at Purdue, my esteemed colleague in all the endeavors that led to and stem from this book, farms and teaches according to the finest principles a reader could find articulated in its pages. James Vorst; Richard Blanton; Charles Boebel; Bob F. Jones, a professor of agricultural economics at Purdue University; and Larry Butler, a professor of biochemistry at Purdue, walked down its lines of prose and culled many a weed. They left only those, as Jesus' parable says, that would pull up the wheat with the darnal.

Introduction

The Argument with Land

As chapter 1 suggests, the root reason why modern agriculture is argumentative is that agriculture itself has always been a human argument with the land. A farmer says "soybeans," and the field answers back "weeds." The farmer contends against the weeds with chemical herbicides. The porosity of the soil responds by leaching a poison into the watershed. A shepherd says "sheep," and the hills echo with "wolves." The shepherd contends against the wolves with poison. Coydogs move into the newly opened territory. They prove too smart to fall for tricked-up bait meat. It is almost as if the shepherd had bred new obstacles to his endeavor.

Today, as always, the farmer and the shepherd must counter again. They cannot afford to let nature get the upper hand in the serious business of producing the foods humans have to eat. But this time, each is likely to have other arguments on his hands from people with strong opinions about how best to control the weeds and wandering pests that infest the fields of agriculture. Some might favor increasing the doses of chemicals and poisons. Others might favor ecological methods that force nature to use her weight and strength against herself. The farmer might try growing companion crops to inhibit the weeds or insects that reduce yields. The shepherd might try an old Navaho trick. According to Hal L. Black and Jeffrey S. Green, writing in the *Journal of Range Management,* the Navahos raise herd dogs that live with their sheep. Roving wild canines shy at the smell of dogs and run off, often with the herd dogs blissfully unaware of their own power.

The vigorous exchange of opinions on these matters often takes place far from the field. Some are scientific. Researchers in laboratories argue with each other at the coffee machine and in journals about the best counterstrategies agriculture should make against nature. Some are ethical. Philosophers might wonder whether pigs, chickens, sheep, or cattle have the same rights humans have to a decent and comfortable life, at least until the day the ax falls. Some are political. Representatives of farmers argue with representatives of city people and consumers. They debate whether farmers ought to be delivering food to the cities as cheaply as possible (usually, but not always, the argument of city folk); or whether farmers ought to be taking special care of their farmland, even if it makes the price of that food more costly (usually, but not always, the arguments of farmers). Some of the political arguments have become dangerous, as when a debt-ridden midwestern farmer became angry enough at the landlord (and banker) to take up a gun.

Even if the arguments remain decorous, passions can still run high. This is probably because there is a lot of money and people's livelihoods involved in farming. But these arguments also belong to a fundamental inquiry into the dependency of human life on an alien nature. As the modern adage puts an ancient piece of wisdom, we are what we eat. When we grow what we need to eat, we impose human demands on nature. Nature resists, and we must try to think of an appropriate or sustainable response. As we do, what we learn from the arguments of agriculture we learn about ourselves, our powers, and our limitations.

Readings

Black, Hal L., and Jeffrey S. Green. "Navaho Use of Mixed-breed Dogs for Management of Predators." *Journal of Range Management* 38, no. 1 (January, 1984).

Chapter 1

The Nature of Agriculture

The Cases

Settling Down

According to its legend, a nomadic tribe has been pasturing its animals in the lowlands during the winter and the uplands during the summer since the beginning of time. The people depend on these animals for most of their food of meat, milk, and blood and their clothing of hides and wool. They grow plots of vegetables during the summer at the upland summer grazing grounds. Although they move their tents and animals at the same time each year during the spring and autumn equinox, they do not return to precisely the same place. They set up their tents adjacent to what looks to be the lushest pastures available when they arrive.

These moving times are times of great excitement. The people dance and feast at resting places along the way. At a solemn moment the tribal elders pray for guidance about where to pitch the tents this season. They believe that while they are in one place, their gods work to prepare the proper pasture in the area to which they will return. The elders must discover the precise location of the new pasture the gods intend. Sometimes the site will command a view of a mountain ridge, sometimes a view of a broad expanse of lake. It is always fruitful. Life has been good with the tribe. Their gods have been kind and have always led them beside green pastures.

At the end of one winter, however, one member of the tribe, restless in a special sense, because he is restless to change from a life of wandering to a life of settling, decides he and his family have moved for the last time. They like the view of the distant mountains from their lowland winter pasture and desire to plant a big garden as soon as the snow leaves in spring, before the spring equinox, and then to stay and cultivate it until harvest, perhaps after the autumn equinox. They decide not to move upland for the summer. They reason that they will have a better garden if they stay put and plant early. They plan to pasture their animals nearby.

The elders are aghast. They believe that the gods will not come to prepare a new pasture in the lowlands for them to return to in the fall if one family stays behind during the summer. The quality of the pasture land will deteriorate. Long arguments ensue. The result is that several other families join the family that wishes to remain behind.

The tribe splits into a farming group and a pasturing group. They remain on friendly terms at first because they still share relatives and beliefs and exchange goods and marriage partners; but gradually their common interests dwindle. One group becomes suspicious of the other. The elders find their worst fears realized. The lowland pastures to which they return in the fall have been heavily grazed and do not sustain their large herds for much of the winter, as in times of old. The lowland farming group becomes resentful of what they think is annual fall intrusion into their land. After all, they reason, they stay behind in the summer to care for it with their animals and their gardens. The pasturing group has no right to demand the use of what they abandon every year. One fall the farming group forbids the pasturing group to return the next spring. There is a battle. The farming group routs the pasturing group. Greatly diminished in numbers, the pasturing group takes to wandering from place to place in strange territories where the gods that prepare good pasture are never able to anticipate their arrival.

Trading Up

An American midwestern farmer lives and works on land that her ancestors wandered into from abroad and settled a hundred years ago. It has been good land, sustaining the health and happiness of four generations. Most of the family food has been grown on the family land. From a seat on the back porch swing one can see orchards, gardens, enclosures for pigs and chickens, and fields with good fences on which generations of farmers have rotated corn, wheat, oats, sheep, and cattle in succeeding seasons. But lately the farmer finds the market price of everything the farm can produce has gone down below the price of producing it, save the price for corn. Even the price of corn is not good, but the local extension agent tells her if a farmer can plant enough of it, the difference in the cost of planting an acre and the slightly higher profit from harvesting it might keep the farm solvent. The farmer uses her personal computer to figure that by planting only corn she can earn in a good year enough cash to maintain payments on her combine, new television, car, and refrigerator, and to save some for the education of her children. The farmer sells her stock, plants her fields and pastures to corn, and begins to tear down the fences between them in order to make one large field on which to maneuver her machines.

Moving On

A young man who was raised on a large grain farm had always wanted to do nothing more than farm himself. But he was the second son in the family, and his older brother had also decided to farm, and if the two of them divided up the family homesteads, each of the two halves would be too small to make a living for each farmer and his family. They talk over their desires with their parents, and they all come to the decision that the parents will retire to Alaska to mine gold for a hobby, the older brother will stay on the farm and will support the younger brother, who will attend a land-grant university. There the younger brother majors in plant genetics. The parents and both brothers do very well.

When it comes time to graduate from college, the younger brother is faced with another decision. He has a good job offer from an agribusiness firm that specializes in developing new hybrids of wheat and corn; he has also been accepted into his university's special graduate program in plant genetics. He could work with a professor he has come to know and respect. This professor believes it might someday be possible to engineer genetically new kinds of perennial grain plants that would be able to fertilize themselves by fixing nitrogen from the air, protect themselves from pests by exuding toxins, and remain deeply rooted in the ground harvest after harvest. There would be no need to disturb the soil around them with plows and chemical rigs, and therefore their cultivation would cause no erosion. The younger brother talks with his family about his opportunities, and they all urge him to accept the job with the agribusiness firm. The research professor's scheme seems to them cockeyed, utopian, a waste of time and energy. If it were possible to design such a plant, his father says, nature would already have done it. Nonetheless the younger brother decides to attend graduate school and work with the visionary professor.

The Issues

When Archimedes once said that if he had a lever big enough he could move the world, he did not really think he would ever find one, of course. His statement was a metaphor that summed up in a succinct image the essence of physics in his time and our own. He was saying that if he knew just the right force to apply at just the right point, anything—even a mountain or a planet—will move. With this knowledge in hand, we could develop techniques for pushing huge amounts of earth around. We would also understand something about the heft and nature of the earth itself.

Perhaps Archimedes also had agriculture in mind. After all, "field-cultivation" or agri-culture is really nothing more than using various levers for pushing the earth around. Draft animals' hooves strike the earth at an angle. A tribal farmer pokes a planting stick into the soil. A tractor pulls a steel plow through sod. It would be useful to be able to stand back at the right place from which we could see what our leverage was doing to the earth.

The Inner-view

The right place would provide two perspectives. We could call one an "inner-view" whereby we look into agriculture's structures and patterns—the continually turning, inner workings of its clock. With the inner-view we see agriculture as a yearly struggle with nature for our food and civilization. Nature inclines to keep plants and animals in ecological balance. Humans want to distort the balance to produce selected plants and animals in one place in sufficient surplus for food and profit. Every advantage human farmers take over nature is only temporary. Nature pushes back with what humans label erosion, pests, or weeds, inciting farmers to yet new strategies of resistance. From this perspective all earthly life as we know it is nothing more than the result of conflict and compromise between these two forces. The basic conflict never changes. As human beings become more ingenious in the ways they devise to beat nature at her game, nature remains resilient in playing it according to her own rules. Today, for example, a modern farmer might have a chemical pesticide at his disposal that kills millions of insects with a single application. The advantage lasts only until the insects evolve resistant strains. The farmer has to try another chemical, or strategy, lest newly numerous and virulent pests chew his farm back to the Stone Age.

People who take the inner-view of agriculture are attracted to terms such as "alternative agriculture," "organic farming," or "sustainable agriculture." We could call them "sustainers." They are inclined to think that it takes ecologically-minded strategies to stay in the game. Human ingenuity must figure out how to get natural fecundity on its side. They are inclined to think that, although the technology of agriculture has changed a great deal in human history, the nature of the conflict at the center of agriculture has not.

Sustainers are inclined as well to a certain moral belief that only healthy yeomen farmers grow healthy food. According to this view, yeomen farmers remain intimate with nature's changes, eat well, get good exercise with hard work, and live in small communities in which individuals greatly depend on one another. They rest assured every night. Families of such farmers are as happy in their own way (and as occasionally afraid and neurotic) as a modern young professional couple working together in New York City as executives in a large multinational corporation. The city-dwelling professionals play tennis for exercise and enjoy frequent nights out in a health-food restaurant filled with a jungle of hanging pots of ferns. But their sophisticated culture, based on modern agriculture, might have forced a wedge between them and a fascinating natural world that their relatives back on the farm love as deeply as the executives love their Porsche.

From this perspective, the history of agriculture may not be progressive. It is rather a process of exchanging one good for another while playing a game whose rules remain the same. A farmer should think long and hard before deciding to give up the advantages of one kind of farming for another. The trade-offs are not always going to be to his or her advantage—or for the long-term benefit of the civilization its farmers feed. For sustainers, agriculture is a zero-sum game. There are no winners and losers over the long term, only players.

The Over-view

If the inner-view looks at the clockworks, the over-view looks at the sweep of the hands over the dial. With it one sees modern agriculture culminating a gradual progression of human activity over time. First the human animal was primarily a hunter-gatherer like other animals, hunting for meat that could be killed and eaten raw, searching in promising places for fresh wild plants and seeds and fruit good to eat. Even today in remote forests, some people still get their food this way. Then in time, certain human groups living near the Tigris and Euphrates rivers (at the setting of the biblical Eden) domesticated plants and animals, while intermittently wandering from place to place when the fertility or grazing land wore out. Even today some nomads farm this way. Then in time, certain groups of humans figured out how to farm "in place." They discovered how to return fertility to the soil or to rotate animal pastures so that they could fence in and use the same pastures year after year. This kind of agriculture goes along with the development of cities and towns because it can set up shop near such settlements and supply them on a permanent basis. Once cities and towns can depend on a steady supply of food, their citizens are assured of the leisure they need to specialize in art, politics, and public works, and to pursue their personal fortunes. As things go on, and as they do in our day, people begin acquiring houses, TVs, automobiles, and sound systems—they can even find themselves competing for resources with their fellows still left on the farm.

People inclined to the over-view are proud of agricultural accomplishments. We could call them "progressives." Agriculture seems to them to culminate in the sophisticated advances of Western culture. Over time and over vast expanses of land, farmers have learned how to raise greater harvests with fewer hands, freeing up more and more individuals to work at building, pottery, politics, and art. Progressives are inclined to the moral belief that farmers should be encouraged to invest in tools and techniques that will make their agriculture more productive. According to this view, farmers should do more with less, continually increasing the disproportion between smaller inputs of labor, seeds, feed, and fertilizer and larger outputs of eggs, meat, fiber, fruit, grain, and vegetables at harvest. Rather than a zero-sum game, agriculture is a gamble human beings can win.

We now turn to a review of some broad perspectives on agriculture to be found in the readings listed at the end of this chapter. They are all inner-views. They offer perspectives on the three cases with which the chapter begins. At this point we need not consider opposing over-views, because even the writers who share inner-views disagree whether it is feasible to put them into practice. Their conflicts are sufficient to occupy us at this point. With the next chapter we begin to consider over-views and eventually consider some writers who try to combine the two.

The Review

Agriculture as Adversary

The agricultural ecologists George W. Cox and Michael D. Atkins offer an analytical inner-view of agriculture in their book *Agricultural Ecology.* They consider nature to be an adversary of agriculture right down at the fundamental level of the genes of plants and animals. In their view, nature tends toward genetic variability, as all the plants, animals, insects, and their diseases in an area adapt to living with each other. They do this by eating each other and escaping being eaten in such a way as to force compromise on their fellow creatures. Basically, no one species can be completely successful in preying on another, lest the one eat itself out of business. The compromise works through the genes of each species. They are continually changing in such a way as to create different subspecies that are more or less susceptible to being eaten at one time and space. The more tasty and thus susceptible subspecies are eaten first, giving time and space to the less tasty and susceptible to take root and thrive, until the preying species itself develops a subspecies with a zest for the new food subspecies by the same process. And so it goes. All the living things in one place continually adapt to each other. Recent studies have shown that these times and spaces are often remarkably small. Creatures adapt to each other quickly "within historical time," with "significant patterns of genetic variation in some species ... along habitat gradients in quite localized situations" (p. 514). This means that in nature there is a considerable degree of "genetic plasticity" in each species that other species can squeeze and shape to their needs while they themselves are being squeezed. The term Cox and Atkins use for the double squeezing is "co-evolution," which leads "toward a state of mutualism, that is, toward conditions that are most favorable to the survival and population stability of all the interacting species" (p. 516). Nature changes things; it tends toward polycultures.

A proto-agriculture begins at a certain point when human beings decide to take advantage of these changes for their own good. At first, as hunter-gatherers, they select from the wide variety in nature the plants and animals most to their taste. In so doing, they act like any other predator species in an area. Ironically, they encourage the survival of subspecies not necessarily to their taste. Such choices, in turn, lead to their wandering from place to place in search of the desired fruits. It also encourages tastes to change and selects for survival those hunters and gatherers most adaptable to change.

Agriculture proper begins the moment human beings begin to take steps to encourage the survival and multiplication of the varieties they find most tasteful. At this point they discover the concept of weed and pest. The weed becomes any plant that hinders the growth of the desired food. The pest becomes any animal that also wants to eat it, from a tiny virus to the big bad wolf. Humans begin to isolate the desired food from its natural competition (the agents of its continual adaptation) and thus to impose "genetic uniformity" onto the artificial landscapes they create for their cultivations. Genetic uniformity becomes "generic vulnerability" as each selected crop offers the pests that love to eat it a completely unintended field day. But the risks are taken with the belief that defenses can be found that will protect the desired crop. In this way agriculture tends to resist the same variety that nature encourages; it tends toward monocultures.

Early agriculture and today's archaic agriculture preserve a lot of natural selection in their workings. Farmers move from place to place as an area ceases to support a desired crop. Or farmers select as new crops varieties nature develops that can resist for a time a ruinous predation. But modern agriculture is different. It features monocultures for several reasons in addition to the major reason that genetically idealized crops can be very productive. Monocultures are also easy to plant and harvest with labor-efficient machines. Consumers come to desire food with a uniform appearance. In Cox's and Atkin's words: "The selection of varieties with desirable inherited characters is inherent in modern agriculture and extreme uniformity arises from this and from the way in which these varieties reproduce or are propagated" (p. 519). With modern agriculture, the payoffs and the risks increase in direct proportion to each other. The payoffs can be large when varieties are successfully bred for large returns on the invested seed and convenience in production and packaging. But Cox and Atkins detail the results of disastrous crop failures such as the Irish potato famine in the 1840s and more recently, the virulent Victoria oat blight of the 1940s and the southern corn leaf blight of the 1960s. These failures were the direct result of large monocultures suddenly becoming susceptible to pests and pathogens rapidly evolving to prey in the fields of the modern farmer (pp. 519–22). The authors also describe the current vulnerability of modern corn and wheat hybrids to rapidly evolving pests and pathogens.

As Cox and Atkins see it, modern agriculture takes over the burden of natural selection as it becomes an ongoing struggle to develop new varieties that will resist continuously evolving pests and pathogens. They write: "In the northwestern United States, the average 'lifetime' of a new variety [of wheat] is about five years" (p. 523). The same basic patterns hold in the breeding and production of livestock (pp. 524–28). This suggests that agriculture is really not the adversary of nature after all, because it is not independent and certainly not an equal. Whatever agriculture appears to win from nature agriculture will eventually lose. Nature always wins the game in the end; and the final score is zero-sum.

This is the conclusion the anthropologist David Rindos comes to in his book *The Origins of Agriculture.* He argues that for every effort humans have made to control their environment with agricultural techniques, they have actually become more dependent on practices that threaten the environment and threaten themselves. The cause of agricultural instability is agriculture itself: "The more effective agricultural procedures are in reducing environmental limitations, the more likely they are to create new opportunities for failure of the system" (p. 275). According to this view, which we could call the adversarial view of agriculture (agriculture versus nature), it is not so much a question whether humans win or lose (agriculture cannot beat nature in the end), but how they can stay in the game. If humans play it well, they can keep agriculture one move ahead of nature's continual countermoves, perhaps indefinitely, or at least as long as human ingenuity and luck hold out.

The Sunny Style

Cox and Atkins, Rindos, and the authors of other readings cited here, notably the ecology-minded I. G. Simmons and Wes Jackson, all have ideas about how humans should play the farming game. They assume we have the luxury of choice among several "styles."

In their chapter "Energy Costs of Agriculture," Cox and Atkins put it simply. Nature grows plants and animals by using the energy of the sun and the water of the rain and by regulating their use by the mechanisms of coevolution. Imagine Mother Nature as the farmer with an invisible hand directing things to grow this way and that. When human beings oppose nature with agriculture, human beings take the regulatory role to themselves, and thereby "must also assume the work involved in many of these regulatory processes" (p. 597). They call this work "energy" and broadly distinguish several kinds. There is ecological energy that comes freely and randomly with the workings of the sun on the earth. There are two kinds of cultural energy humans supply to regulate ecological energy to their special advantage. Biological energy comes from their labor and the labor and manures of their animals; industrial energy comes from their chemicals and machines.

They refine these distinctions. There are three basic systems of non-mechanized or biological labor: "pastoral systems; cropping systems that rely exclusively on human labor; and cropping systems that use draft animals" (p. 598). They make these distinctions very interesting when they begin to calculate the amount of energy humans expend in each of these systems in proportion to the amount of energy they reap in the calories of the food they harvest. They say that there is little data measuring the ratios of input to output for strictly pastoral farming; but what data there are suggest that for every calorie of energy a family expends in herding its animals, they reap 9.7 more, or a very healthy return. They can do this because by moving their animals from place to place they take advantage of the "free" energy nature expends on scattered wild pastures. More data are available on the energy costs of semi-pastoral farming, where farmers tend both animals and vegetable crops, periodically move their animals between pasturing grounds, and feed some of their crops to their animals as a method of storing up surplus and fat against possibly lean times. (This is the early farming of the tribe in the first case of this chapter.) In the most primitive form of this system of farming, the energy ratio is lower, approximately one calorie expended for three calories of energy earned. But with the use of simple tools the energy balance becomes more positive, perhaps reaching proportions of one to forty. With the additions of the use of draft animals for ground cultivation and irrigation systems exploiting local water supplies and seasonal flooding the proportions remain about the same, or perhaps become a little less. But now farmers can trade one good for another: the good of being able to stay in one place for a slight loss in energy paybacks; and there is considerable room for manipulation. If the farmers put basic industrial energy into developing sophisticated plows and pest control and harnassing more animals to their field labor, they begin to discover ways of transforming inedible food into edible food by the simple process of feeding things to their animals that they cannot eat, such as stalks, grass, and peels, and using the resulting vigor of their animals to grow edible food. In some farming societies of India that use a modified draft animal, primitive industrial system, the overall energy ratio is lower than in the semipastoral system, but the ratio between industrial energy input and food calorie output can be as high as one to three hundred (pp. 598–602). (See chart 1.1.) In these systems humans integrate themselves into natural systems and manipulate them to their advantage and thus seem to stay ahead of nature purely and simply. The trick is that their human labor becomes a part of natural energy, and thus humans win the game by joining the other side. (The reader is referred to the readings listed at the end of this chapter for the quantitative studies supporting the general conclusions reviewed here.)

CHART 1.1   Ratios between Calories of Energy Input to Food Calorie Output

Only solar and cultural energy used

Nomadic herding	1:	9.7
Semipastoral farming without tools	1:	3.0
Semipastoral farming with tools	1:	40.0

Industrial energy added

Modified draft animal farming	1:	300.0
Industrial farming of beef	1:	.2
Industrial farming of pork	1:	.6
Industrial farming in general	1:	.102

SOURCE: Cox and Atkins

The Oily Style

These ratios change drastically as agriculture mechanizes. Now the proportion between the calories of human energy expended by one farmer sitting on a tractor and the amount of the calories of food his fields produce can become astronomical—one to many thousands—but the overall proportion plunges to the negative when all the energy costs are factored in. They include the fossil-fuel energy it takes to run the farm machinery, the energy it takes to make the tractor and its implements in the first place, as well as to process seeds and manufacture chemicals. Perhaps we should include the amount of energy it takes to educate the farmer to use all of today's modern sophisticated farming gear. Cox and Atkins admit that it is difficult to measure these ratios precisely, because energy comes in different forms with different concentrations of usefulness. "For example, one kilocalorie of solar radiation has much less capacity to do useful work than a kilocalorie of energy in the form of electricity, in fact, it has approximately 2000 times less" (p. 602). But when the proper conversions are made, the end result is negative. Once again farmers convert something humans cannot eat, in this case petroleum instead of stalks, into something they can, such as grain or meat, but with considerably diminishing returns. The authors cite several studies and supply various charts that point to the same conclusion: "These analyses show that most vegetable crops and fruits yield food energy in amounts equal to or less than the cultural energy investment" (p. 607).

When these crops are fed to animals and food is harvested in the form of meat, the ratio plunges to as little as 0.2 for beef and 0.6 for pork and goes lower when the costs and energy for processing and marketing and perhaps even charcoal grilling are factored in. (See chart 1:1.) Now the trade-off is simply great amounts of energy for the convenience and taste of butchered, sizzling meat: "the average industrial energy subsidy for animal food protein is about 30 to 35 calories per food calorie" (p. 613). It is about thirty calories for fish. I. G. Simmons, in his book *The Ecology of Natural Resources,* makes a similar case and offers the astonishing calculation that it takes 132.5 liters of water to produce a single slice of white bread (p. 197). Cox and Atkins cite an overall estimate of the ratio for all food produced in this country as 9.8 calories of cultural energy per calorie of food consumed.

Various Regimens for Health

From these calculations it appears that modern agriculture is on an energy binge; and as there is only so much energy available, that binge will have to come to an end sooner or later. Now, one could argue that these arguments are basically saying that wine is far more expensive than well water and is a lot less good for you. Well, perhaps. But wine tastes a lot better than water, and there are certain very important occasions when the pleasure it imparts to a meal cannot be surpassed. Or, to put this in drier terms, the fact that a product in one form demands far more kilocalories to produce than a previous form did seems to be irrelevant if the new form has a higher utility value to the user.

But the sort of radical analysis that Cox, Atkins, and Simmons provide gives a clear picture of the hidden but true costs of our modern eating habits. It is a useful picture to have for understanding the many fundamental arguments of agriculture, even if one does not come to the same conclusions these three writers reach about stopping the binge. They themselves are interested in preventive medicine.

For Cox and Atkins the watchwords are "diversification" and "efficiency." Interestingly, when we consider what they mean by each term, their overall ethics for agriculture strike a balance between innovation and preservation. For example, when they stress the importance of genetic conservation for agriculture, they are talking about very modern laboratories working with very ancient and natural genetic materials. They praise the efforts of many governments to set up genetic banks where different types of seed from different varieties of domestic and wild plants can be stored against a future need for new genetic material. The idea is that nature is sure to resist our monocultures with pests and pathogens as yet unknown and that we need all the resources we can gather to muster the appropriate counterattack (pp. 528–31). They reason that such banks need to be expanded wherever possible and that new studies need to be done on the data their genetic museum can supply on the mechanisms nature uses to evolve natural defenses against new pests and pathogens: "The need to make rapid responses in breeding for resistance is growing, and immediate access to information is critical" (p. 533). They also suggest that agriculture mimic nature "by diversifying agroecosystems in terms of crops and varieties so that all the eggs are not in one basket" (p. 534). When they take up questions of future energy use, they emphasize even more the need to become aware of the methods of an older and more diversified agriculture. Beef could be raised more cheaply, and its meat could be more healthy to eat when grass-fed on large pastures. Human labor could substitute for machine labor and perhaps be more careful in cultivating and harvesting fruit. Rotating crops with animals on the same fields in different years could reduce the need for artificial fertilizer whose production is very energy intensive. More farming could use organic methods. A study of different cultivation methods used by different Amish communities suggests that the cultural energy input can dip below the calorie energy harvest on a wider scale than the Amish customarily farm (p. 625). In general what we need is a scientific acuteness and dispassionate interest in the best in traditional and modern agriculture, with one eye cocked to the energy input/output balance sheet.

I. G. Simmons comes to many of the same conclusions. His watchword is "intensification." Nations need to learn to farm more areas of their arable land more efficiently, and this can be done primarily by educating the farmers already on the land about what agricultural scientists and technicians have already learned. At the same time, agricultural intellectuals need to learn all they can about the culture of the farmers they hope to educate and make more efficient. In India, for example, during what has been called "the Green Revolution," new varieties of grain greatly increased the productivity of the farmland immediately. But the higher yields came at a price. The new techniques required larger farms and fewer farmers to tend them. As a result many people were forced off the land without any prospects of jobs in the cities and industry to give them a livelihood (p. 189).

With these writers' hopes in mind, it is an interesting question to turn to Wes Jackson's book *New Roots for Agriculture* and ask if his specific program for reviving a more natural agriculture is as sound as it is succinct. More importantly, is it feasible?

His succinct summary of the tension between agriculture and nature is in line with others we have read: "To maintain the 'ever-normal' granary the agricultural human's pull historically has been toward the monoculture of annuals. Nature's pull is toward a polyculture of perennials" (p. 93). He uses the image of a bill collector to extend the contrast. Whenever a living thing in nature temporarily prospers at the expense of its neighbor or predator, nature eventually presents a bill in the form of a plague or predation that both cancels some of the gain and forces the survivors to adapt to new conditions. Whenever human farmers force their beloved plants and animals to prosper at the expense of their natural competitors or predators, they try to fend the natural bill collector off with their harsh chemicals and spiky implements. Eventually, however, Jackson claims, the bills must be paid. He says that although humans might think that they can come up with new pesticides and herbicides and tools to hold nature at bay indefinitely, even today the bills are being collected. While modern agriculture keeps an eye on the creatures trying to sneak over the fences, nature is pulling the soil out from under its feet in erosion. Modern agriculture, he says, is dissipating its living blood: "The soil is going fast" (p. 99). His images and writing are more vivid than the other authors we have already considered, but like the others he thinks that agriculture should become more natural. Farmers should spread out their payments instead of paying all the bills at once. "If we are to heal the split, it is the human agricultural system which must grow more toward the ways of nature than the other way around" (p. 94).

Jackson's plan for how our agriculture might become more natural does seem to be more radical than the others discussed in this chapter in two ways. On the one hand, it is more archaic. He sees the ideal modern agriculturalist trying to think up clever ways to go back to primitive agricultural styles that used less fossil fuel energy and chemical fertilizer and fewer pesticides than modern agriculture does today. And then he wants to go back even further to an agriculture that resembles the state of a healthy nature herself. His new perennial pastures would provide "a crop [that] could go a long way toward preventing erosion and desertification . . ." He thus identifies his desire to turn back the clock to an earlier time when a healthier landscape flourished (p. 102).

On the other hand, Jackson's scenario is more futuristic. In the future our agriculture should come to rely on the harvesting of perennial grain crops that do not exist right now. They have yet to be discovered or even engineered. At Jackson's Land Institute, he and his fellow researchers and students are banking both domestic and wild seed and plant varieties from all over the world. They are doing more than store them, however. They are experimenting with developing a perennial grain crop that remains rooted in the ground for several or many years. It would thus inhibit erosion. They hope that their cross-breeding might also produce grain crops with yields per acre close to the seed-rich annuals upon which our grain agriculture now depends. Certain experiments they have already done suggest that such yields are theoretically possible, even now. To become commercially feasible, more work needs to be done. They need to develop not only new plants but also new ways to plant and grow them together. Companion plantings might allow plants to protect each other while they mature. Their maturity would need to be synchronized so that they could be harvested together by machines. Granted, Jackson says, these new plants might require a change in human tastes. We would have to learn to like breads made from new grains and give up our dependence on refined wheat flour. And this is a large task. But tastes have changed before and can again. Jackson is optimistic: "If only a hundreth of the advertising is applied to the promotion of eating these healthful grains that is applied to the array of unwholesome junk food we ingest now, no culture barrier can stand in the way of their whole-hearted adoption" (p. 111). Then the future grain field in the Midwest might come to resemble the mixed grass perennial prairie that once covered it.

When Jackson says that the future natural farm would be the product of "holistic thinking" (p. 105), he is developing a vision of the future of agriculture that goes beyond scientific feasibility to utopian hopefulness. Like all utopias, the one he describes at the end of his book has to be set in the future. That is the only place where it seems possible in the present to recover the real values of the past. Jackson's agricultural past is very far back, certainly much older than any of the toilsome farming done in this country during the four centuries of her colonization and settlement. In essence, it is the past of native farmers, now brought up-to-date with central heating and interior plumbing. Jackson envisions a future "front porch farming" where the plants grow naturally. They fertilize and protect themselves in communities of plants much akin to the mixed grass perennial prairie that once covered the midwestern United States. The farmer recovers the leisure anthropologists attribute to primitive farmers. They needed only several hours of earnest work a day to earn their daily bread. Like them, the future farmers of America will have ample time for basket weaving and dancing (pp. 112–13).

Conclusion

An intriguing statement of Wes Jackson provides an interesting perspective on the material in this chapter: "It has yet to sink into our culture that we are still basically gatherers and hunters, and that the era of agriculture is but a thin veneer over an evolutionary past that tolerated a great deal of leisure" (p. 112). He foresees a time when farmers might once again enjoy the leisure of their remotest ancestors. He implies in this context and other places in his final chapters that one of the reasons agriculture might have developed the way it has is out of a perverse human desire to make things hard for himself. So-called labor-saving devices only make life harder in the long run by making it possible for humans to do more work than they were ever able to do with their hands and simple tools.

These are interesting ideas in their own right. But we can take this statement and use it as a summary for an inquiry into the nature of agriculture. Perhaps we have never gotten over the fundamental beliefs of our hunter-gatherer ancestors. We still believe someplace deep down in our common pysche that we can stay one step ahead of nature by simply moving on to another environment, another landscape. Until one hundred years ago this belief still had some validity in this country, as farmers could migrate from exhausted land in the East to the virgin prairies of the Middle West. But the belief persists in this country and throughout the rest of the settled modern world. Now it is transferred to the process of scientific discovery which permits a transformation of land in place of a migration of the farmer. That is to say, if farms and farmers now must stay put, science and technology will permit them a kind of artificial migration from traditional farming to modern farming whereby they change the same fields into an entirely different landscape. The split between nature and agriculture that Wes Jackson deplores must only grow wider.

David Rindos, an agricultural historian, shares many of these views. In a very densely argued and interesting chapter in his book on the history of agriculture, he defines the nature of agriculture in a phrase to which our other writers would subscribe: "the two major effects of agriculture are an increase in productivity and a concomitant increase in the instability of that productivity" (p. 276). But he comes close to despair when he considers whether we humans are capable of making such a revolution of thought as Jackson describes. We would need to make a conscious decision as a species, which we have never done. For Rindos, much of human cultural evolution has been as unconscious and provisional as natural evolution. Our ancestors stopped nomadic farming and started settled farming without ever really understanding what the consequences would be, in the same way that an animal breeder cannot predict what the breed will look like in a century. This idea is more than a special emphasis on the fundamental ignorance of human agriculture. It suggests that our case is somewhat hopeless. We never know what we are doing and especially what the consequences might be when we head off in a new direction.

Thus it is somewhat ironic, if not contradictory, when Rindos concludes that as we moderns know more about agriculture and nature than our ancestors did, we are the first generation who must make a conscious decision. We either make the right choices or face disaster. We no longer have the luxury of emigrating away from our agricultural mistakes. At the end he makes the same argument that the others do: "I would hope that an awareness of the processes by which agriculture developed may act as a spur to us to gather the information that may permit us to become as successfully intentional as we have so glibly claimed to be" (p. 285).

All together these voices counsel us to do better—more intensely, consciously—what we are already doing. None suggest we turn back the clock to an older kind of farming. But all suggest we revive the ancient wisdom with which that older farming was done.

In the next chapter we begin to hear the voice of progressives who are much more satisfied with the way things are progressing on the modern farm.

Readings

Cox, George W., and Michael D. Atkins. "Genetic Vulnerability and Germ Plasm Resources"; "Energy Costs of Agriculture." Chaps. 21 and 24 in *Agricultural Ecology.* San Francisco: W. H. Freeman and Company, 1979.

Jackson, Wes. "New Roots for Agriculture"; "Outside the Solar Village: One Utopian Farm." Chaps. 8 and 9 in *New Roots for Agriculture.* Lincoln: University of Nebraska Press, 1985.

Rindos, David. "Agriculture and the Paradigm of Consciousness"; "Instability, Cultural Fecundity, and Dispersals." Chaps. 1 and 6 in *The Origins of Agriculture: an Evolutionary Perspective.* Orlando, Fla.: Academic Press, 1984.

Simmons, I. G. "Food and Agriculture." Chap. 7 in *The Ecology of Natural Resources.* New York: John Wiley and Sons, 1981.

Chapter 2

The Principles of Agriculture

The Cases

A Successful Roman

In the early days of the Roman Empire, Felix Agricola, the oldest of six brothers, inherits from his father a small farm that has been in the family for six generations. It dates back to the times when the first Roman settlements were established on the Italian Peninsula in the gentle hills where the old Sabines used to live. There are walnut groves on the tops of the hills, where legend has it the spirits of the Sabines still roam. Partly out of superstition, partly out of the pleasure their view provides, the Roman farmers have left the groves uncut and even kept their hogs from foraging the walnuts. The Agricola farm lies only fifty leagues from Rome. Although it is not far from the seaport Ora Maris, with good facilities for shipping grain to the capital city, the family farmers who have worked it have disposed of surpluses in the nearby towns. Mostly they have bartered farm goods for tools, clothing, and salt. The bulk of the farm produce has always gone to feed the family, which has always eaten well.

Although small, only about seven *jugera* or four acres, the farm has been intensively productive. There are orchards with six kinds of fruit trees, pens for hogs and sheep, a tiny pasture for a few milk cows, small fields with grain and vegetable plots, and an excellent vineyard. An invitation to an Agricola winetasting party deserves to be printed on parchment, so the local saying goes.

Felix, however, has seen a bit of the world before taking over the farm and thinks of expanding the operation. He fought in the Second Punic War, rising to the rank of Centurian when he distinguished himself in battle. He spent time in the city of Rome before being mustered out. There he became acquainted with some grain merchants who were the fathers of some of his former comrades-in-arms. He learned about their plans to begin importing grain in huge refitted warships from the not-so-distant shores of northern Africa.

He returns to the family farm with his war bonus. It is a deed which gives him title to six hundred *jugera* of public land in the adjacent countryside. The big tract of land gives him big ideas. Shortly after arriving home, he calls his brothers together on the homestead farm and tells

19

them his ideas. They like what they hear. They decide to divide the original farmstead into six homesteads, one for each of their families, and to buy a good-size herd of cattle on credit. Felix's meritorious war service also entitles him to borrow money from the government. In addition, they buy a gang of war slaves. They proceed to train some of the slaves to tend the herds and others to tend the grain fields. They plan to pasture animals until their maturity, then fatten them on their own grain before taking them to market.

The brothers become the new farm's managers. They work hard and work their slaves hard. After five years they ship two boat loads of fat cattle to the Roman market. With the profits they buy their own ships. After two more profitable years, they buy more ships. They expand their enterprise into transporting the foodstuffs of other farmers to Rome and beyond. Several of them buy villas on the outskirts of Rome and return to the original homestead on holidays. Felix, a popular businessman, eventually becomes a member of the Roman Senate.

A Clever Peasant

Melchior Pampas grows wonderful corn on one hundred and fifty acres of good, well-watered bottom land near Guadalajara, Mexico. His government has a policy of keeping down the market price of corn in order to keep down the price of corn tortillas—the staple of the poor Mexican's diet. This policy also encourages a migration of small farmers to the country's industrial centers to find more lucrative work. But Melchior has stayed on the farm and prospered. He has been lucky. A tortilla factory has been built nearby. It costs him little to get his corn to this market. So his tiny profit margin has been enough to enable him to replace his adobe hut with a three-room concrete house with electricity and running water.

But now even more prosperity looms. During a winter course at a nearby government agricultural extension station, Melchior learns about sorghum. The new hybrids developed in the United States are very productive and much easier to grow than corn. What is more, the market price is not controlled by the government, and right now that price is very high. There is a big demand for sorghum to add to animal feed to fatten cattle for the increasing hunger for meat in Mexico's expanding cities. In the next growing season, Melchior switches to sorghum. He joins a small revolution in the Mexican countryside. From 1958 to 1980, production of sorghum grows 2,772 percent and the amount of land sown in sorghum climbs 1,300 percent. The number of large-scale livestock operations in Mexico soars; more Mexicans are eating meat, including Melchior and his family; and soon he can afford a car and a TV.

But there is a big catch. While many poor Mexicans continue to eat corn, they cannot eat sorghum and cannot afford to eat meat. The nation's hungry have less food than they would otherwise, and Mexico is forced to spend precious foreign exchange on corn imports for the first time in its history.

The government decides something must be done. It raises the price it will allow to be paid for corn. It cannot raise it very high, lest the price of tortillas rise beyond the means of the urban poor. Therefore extension agents are instructed to explain the situation to their local farmers. They are to encourage the farmers to switch back to corn voluntarily, for the good of the country. Melchior considers the good of the country, then the good of his family, and decides to stay with sorghum.

Back to the Land

Christine is the daughter of a successful Washington C.P.A. She is accepted into Harvard Law School and is elected to the Law Review in her second year. She graduates with honors and, with her father's contacts and her record and obvious intelligence, lands a good job in a New York City law firm. After several years with the firm, she is close to being offered a partnership, and also close to a nervous breakdown. She takes an extended leave of absence during the summer months, which she spends at her family's summer home on an old farmstead in Vermont. As she recuperates, she finds herself getting interested in the life of the local countryside by reading the local paper. She buys milk, eggs, and produce from some of the local families still doing subsistence farming in the hills (usually supporting themselves and their farms with jobs in resorts) and gets to know some of them by name. She finds she likes their directness. Partly for therapy, partly for fun, she puts in a late summer garden behind the house. It is a great success.

On Labor Day she resigns her position in the law firm and with her parents' permission stays on in the farmhouse for the fall. She decides she wants to try farming herself. She visits with the local extension agent, and they work out a plan for planting the old farmstead's pastures to strawberries and raspberries. The extension agent tells her that the soil and the climate favor berries, and that there is a good summertime "u-pick" business for berries beginning in the area, and that a local monastery buys berries for making jam. She studies farming with the same zeal she once studied law, puts in vines and rows, and after several years is producing good, healthy crops using animal manures for fertilizer and cultivating with an old tractor.

Eventually she decides she wants to settle in the area permanently. She calculates her financial situation and decides that she will never be able to make the berry farm pay a profit, even if she could own the land outright. The costs of production are very close to the market price for the produce in an average year and higher in a bad one. Nonetheless, she figures she can afford to live where she loves to live. She inherited a big block of stock in a major weapons contractor from her grandfather; the dividends from the stock will provide her with a good income, whatever money the berries might provide. With her savings she buys the farmstead from her parents, winterizes it, adds a solar addition, and settles in.

Eventually she meets and marries a penniless young physician who has also moved into the hills from the city. He works part-time at a local clinic and expands the farming operation to include angora goats, whose wool he looms into elegant rugs.

The Issues

Two Principles

There are two principles of agriculture. In the abstract, they are very simple. According to one, a farmer farms for the good life. Hard work, clear air, cool water from a well, wholesome food, and a well-run farm provide the farmer and family with a healthy place to live. Profitability is secondary, although still important for buying more worldly goods. According to the other principle, the farmer farms strictly for the money cash crops bring. The farm might be a nice place to work, but it hardly begins to produce all the things a farm family needs to live. Ideally both principles work in tandem, one right after the other, even if on one farm one principle has priority, and on another the other. But in practice the mix can be very complicated.

The complications arise because farming according to either principle, and especially the second, means taking part in a national and global economic community far larger than any community of farmers. This larger community has many priorities of its own. One of them, of course, is selling products like farm equipment, pickup trucks, and television sets to farmers. For this reason, it is important to the larger economic community that farmers get good prices for the food they grow and become as wealthy as any other group of consumers. But farmers are also producers. And in this role farmers often find that the larger economic community is not as friendly. Now it wants the cheap food prices that allow far greater numbers of suburban and urban citizens to spend less of their income for staples and more for luxuries. Cheap food also allows commodity traders to buy farm products at low prices for sale or trade abroad. Farmers often complain that they are expected to buy retail but sell wholesale. As Indiana farmer John Hahn once said, "Farmers pay the price that's asked, and take the price that's given."

With its interest in cheap food, it is not surprising that the larger community has developed methods for encouraging farmers to grow more food more cheaply. One very powerful method has proved to be the land-grant university system. Its ancillary constellations of extension agents keep farmers informed about the latest techniques of ever more productive farming. Companies that sell farm machinery and chemicals advertise their wares usually with promises that farmers will be able to grow more and increase profits.

But increased productivity cuts with a double-edged sickle. In our country, greater productivity has led to a surplus of many commodities. A farm surplus means lower prices per unit produced. Lower prices often require larger-scale productions, requiring more academic research, bigger farms and machines, and more chemicals. Thus the agrieducational-agribusiness complex encourages a continuing spiral of technological investment by farmers of which it reaps the first harvest.

To the extent that farmers are part of the larger economic community then, they are going to find some resistance to their farming according to either principle successfully. This is especially true in a more or less free-market economy. Competition among peers can be sharp. Pressure from forces outside the farm can be heavy, even for the most single-minded money farmer. But to the extent farmers farm first of all to preserve the health of their land and their family, they are going to be more vulnerable to both competition and economic pressure. This principle is virtually a private value of the sort that is always primarily of interest to the valuer. It has to do with pride, a sense of stewardship and craftsmanship, and a desire for a special environment in which to live and raise a family. A value is a value precisely because it cannot be priced. But in this case, one can only hold it at a price. Certain farmers have to decide on their own to invest in the long-term health of the farm—often by sacrificing vital short-term returns.

It is true that there are members of the larger economic community who understand the value of the first principle. They know that farmers who abide by it take good care of the land. Many of their children have left the farm to work in agrieducation and agribusiness. They understand the bad economic consequences of the diminishing numbers of family farms. After all, many segments of the farm economy need farmers to whom to sell seeds, tools, and credit. Some agronomists have become alarmed at the steadily increasing erosion of topsoil that is often the consequence of the harsh field practices that modern-day, large-scale, productive, chemical-industrial farming entails. But they themselves belong to larger institutions and finally to the larger economic community that, like the Cyclops, is huge with a single eye for profit and progress.

Perhaps this is why the farms whose priority is the first principle have a strong hold on our imagination. We need potent antidotes to economic necessity. Picture a large white farmhouse with an inviting porch and rocking chairs. There is a red barn nearby, contented animals grazing under shade trees in pastures enclosed by wooden fences. There is the farmer, smiling, loading the grain hopper from an overflowing cart, while his younger children play with a puppy, and his adolescent children help him work the machinery. Mother watches from the porch, smiling, drying her hands on her apron. This is the farm that is a wholesome and a prosperous place to live. On it hard work pays off. There are no clocks to punch because really none of its workers ever goes off duty. They will rise at midnight to quiet animals during a storm and rise at dawn to cook breakfast. Almost endless variations of this image have appeared on calendars and popular art throughout the history of this country, only the style of clothing and the technical sophistication of the machinery changing from decade to decade, century to century. Its hold on our imagination derives in part from the myth of the new Eden many of us carry in our souls. We believe that now that we must labor by the sweat of our brows, it is only on such a farm that human beings can live in harmony with nature and with themselves outside of paradise. The farm is a place of many living things, humans, animals, plants, and even wild things making up a community that quarrels at times, to be sure, but has worked out a basic accommodation of each other's needs. Even if not all of the members of a culture can afford to live on such a farm, the belief that many of their fellow countrymen still do assures them that even as they enjoy the sophistications of the city, somebody is back home, keeping an eye on the farm and in touch with the fundamentals of life.

Something like this is the image most of us carry in our heads for other more obvious reasons. Many of us are only several years or generations removed from such a place—at least in our fond imaginations; and ours is not the only culture that has honored the image of the doubly principled farm. It abounds in the art and literature of other cultures today and has all through time. The Old Testament speaks of the gift of good land flowing with milk and honey that God will give his chosen people if they remain faithful to him. Homer's hero Odysseus plans the battle to win back his farmstead from the suitors in the humble but honest hut of his family's swineherd and meets success in the orchard his ancient father has kept cultivated until the day of his return. The Roman poet Virgil pictures the owner of a small farm passing a bowl of wine to his laborers seated with him around a camp fire after the harvest. He calls them his friends; and while they celebrate, his animals frolic and his children play games with bows and arrows.

But when we look to the agricultural history of any culture, ancient or modern, we find that the ideal farm rarely or only briefly exists. Real farms are more often farms of desperation where life is hard. At first life is hard because the soil has to be broken and the wilderness just outside the fences has to be tamed. Often native inhabitants such as Indians or Sabines have to be forcibly removed at the price of bloodshed and retaliation. Every improvement the farmers make, wild animals, storms, insects, wars, taxes, or the greed of outsiders threaten. Even well-established farms, enjoying the twin principles of farming, cannot remain intact and withdrawn behind their fences. There are larger and more subtle enemies than locusts and hail. Consumers desire cheap food, and market forces put pressure on farmers to grow bigger surpluses of cheaper food on larger-scale farms. Sometimes these pressures have forced small farmers off the land and forced the consolidation of small farms into bigger ones.

Because the doubly principled farm has a hold on our imagination and seems a good tradition to uphold, and because such farms are hard to establish and hard to hold together once they are made, it is not surprising to find that a lot of farmers and the friends of farmers have thought about the two principles of farming. They have wondered if they might be balanced for the good of the farmer, the farm, and the culture that depends on both for food, fiber, and the leisure to pursue the secondary necessities of art, politics, and trade. We are going to sample only a few of these thinkers here, just enough to give an impression of the extent and breadth of their thought.

The Review

An Ancient Debate

In a recent article in *Agriculture and Human Values,* Jan Wojcik relates how two thousand years ago, about 40 B.C., an eighty-year-old Roman farmer and writer, Marcus Terentius Varro, published a treatise *De Re Rustica,* "On Rural Matters." He wanted his family to use it as a manual on farming when they inherited his farm. In the opening pages he describes himself standing around with some fellow farmers, waiting for a priest to show up to open the doors of the Temple of Tellus, the Goddess of the Earth. It is the first day of the spring sowing festival, and they are all looking forward to a reverent moment of ritual hopefulness before a few days of serious celebrating with wine and cakes and then a long season of farm work. While the farmers gossip and boast and cadge advice from one another just like the farmers today would around the Feed and Seed store, two of them get to arguing about why a farmer should farm. "You need to make some money," says one Gaius Fundanius, whose name translates "Down to Earth," "only so that you can afford to live on a 'wholesome' (*saluber*) piece of ground."

"Nonsense," snorts Gnaeus Tremelus Scrofa, whose name translates loosely "Genuine Quivering Swine," "the only reason one should spend any time making a farm look wholesome is that it might add to the selling price.... A farmer farms," he says, "to make money" (I, ii, 8; I, iv, 3).

Wojcik makes the point that virtually the same arguments can be heard in our day and place. During an interview, David Schwamberger, a young bachelor farmer in a seed company cap and a striped shirt stood up to his knees in a thousand acres of soybeans on an early September day a few years ago, outside of the town of Brookston, Indiana, near the turnoff by the Orville Redenbacher Popcorn silos. He was trying to explain to someone why he farmed. It was his grandmother's old place, he said; he loved Mother Nature; he was handy with tools; he liked working with his dad; the plants he grew looked good on a clear sunny day. He waved his hands, picking reasons out of the air. Then he paused for a moment for emphasis, smiled and said: "Well, I know one thing for sure. You don't farm for the money. There's an old joke that if you gave a farmer a million dollars he'd just keep farming until the money ran out." There was a faint ripple in the leaves ... yellow light smiling on the green ... it sounded like the rustle of money.

At about the same time, Bryan Jones, another young farmer who raises cattle and feed on the high plains in Nebraska, sat with other farmers in a meeting hall in Des Moines, Iowa, surrounded by a passel of government bureaucrats. The bureaucrats wanted to know, among other things, why farmers farm. One said it was to pass on good farmland to the children; another said he wanted to set a good example by farming organically. There was a silence, and then Jones said he farmed for the money and wanted to make lots of it. His fellow farmers booed.

Both Schwamberger and Jones were being facetious, of course. Schwamberger has the *Wall Street Journal* sent directly to his farm; and later Jones got to wondering why cattlemen like himself continue to raise cattle when the industry has made money only fifteen years out of the last one hundred. "Maybe it's all those trips out to the pasture in the spring, seeing little calves hiding in the grass, or butting one another, or running down the hill trying their new legs out." (Bryan Jones, *The Farming Game* [Lincoln: The University of Nebraska Press, 1982].) Both give mischievous answers because the question is so old and cantankerous, answering it is like kidding with your grandparents.

But the question can still provoke serious and carefully reasoned arguments as well, such as we find in selected chapters of *The Unsettling of America* by Wendell Berry and *Transforming Traditional Agriculture* by Theodore W. Schultz. These writers defend the principle of the wholesome farm and the principle of the prosperous farm respectively, and few writers understand them better.

The Wholesome Farm

Wendell Berry is a modern-day Virgil who farms a small diversified farm near Port Royal, Kentucky, and writes poetry in praise of good farming, hard work, and the pleasures of honest community life. He is also an essayist of great power. His extended essay *The Unsettling of America* published in the late 1970s is still considered one of the most eloquent and passionate defenses of wholesome farming in our literature. But it is more than that. It is a broad and incisive argument against exploiters and a defense of sustainers throughout our culture. Farms stand out as the most critical battleground between these two contending parties. The image of the ideal American farm Berry presents is not far from the popular image sketched above; but his essay does not so much praise such farms as lament their disappearance from the American scene. They disappear as the principle of the wholesome farm loses ground to the market principle that the American farm must produce an ever greater abundance of ever cheaper food.

Berry blames politicians for the disappearance of the wholesome farm in favor of a prosperous farm. He frequently cites and ridicules the policy statements of former Secretary of Agriculture Earl Butz and his supporters, all of which are variations on Butz's central theme that modern American farmers have to get big or get out. According to them, it is a blessing that fewer and fewer farmers grow a bigger and bigger percentage of our food, thus freeing the rest of us for more interesting and lucrative occupations off the farm.

But Berry thinks that the popularity of Butz's basic ideas is broader and more enduring than any one person's articulation of them. He began thinking about writing his essay, he says, in response to a 1967 report of President Johnson's special commission on federal food and fiber policies. According to this commission the country's biggest farm problem was a surplus of farmers: "the technological advances in agriculture have so greatly reduced the need for manpower that too many people are trying to live on a national farm income wholly inadequate for them." Berry found himself so astonished to read that the government was valuing "technological advances" more than the lives and communities of small farmers still on the land. "Reading that [report] I realized that my values were not only out of fashion, but under powerful attack. I saw that I was a member of a threatened minority. That is what set me off" (p. viii).

Berry then develops an argument that the conflict between Butz's basic ideas and his old-fashioned values originates at a level deeper than a disagreement between politicians and farmers or consumers and farmers. It originates in the American character itself. "There is no use pretending that the contradiction between what we think or say and what we do is a limited phenomenon," he says, speaking of the tendency we have to hold certain ideals inconsistent with the way we actually live our lives (p. 18). He says this in discussing, first of all, the embarrassing discovery the Sierra Club made that it owned stock in companies whose industrial practices threatened the very wildlands the Sierra Club was committed to preserving. This is a discovery we can all make about ourselves, he goes on to say, especially when what he calls the "disease" of specialization blinds many individuals within our society to have only a

very narrow understanding of how things work. They come to understand only the demands of their discipline, not how that discipline inter-
acts with others, and are content to be well paid for its practice and to
remain oblivious to its consequences. In the process "workmanship,
care, conscience and responsibility" go by the board and are scarcely
missed. Most of us nowadays go through an entire day without ever
touching anything primary. We know nothing about the soil that produces our food, let alone about the poisons that have been used to kill
its pests or the salaries of the workers who have been pressed into its
harvest. Few of us satisfy any of our basic needs for clothing or shelter
or see any raw material become a finished product or even understand
how it is done.

As a consequence, Berry believes, most modern Americans are deeply
unhappy. They rely on other specialists they do not know or care about
to satisfy their own basic needs; and as those specialists care nothing for
the consumer or for the products they produce, most of us wind up
eating unhealthy food, own mass-produced autos and houses that are
continually breaking down, breathe dirty air, and drink tainted water.
Berry believes that the only formula for happiness is deciding to do
things well oneself. Turning away from the world means turning into
the self; in this way, many individuals who have come to insist on quality
for themselves will be able to cultivate a better society at the grass roots.

Berry desires a simple revolution, whereby a homemaker would insist
on quality ingredients in the food he or she prepares directly out of
basic foodstuffs "with some intelligent regard for the nutritional value"
(p. 24). This homemaker would buy only good-quality raw materials
from the grocer for meals prepared mainly in the home. Such careful
shopping requires the grocer to stock the quality raw materials the
homemaker needs, and in turn, the farmer who supplies the grocer to
grow the good-quality raw materials that will sell. Because these materials would require careful farmers to grow them, those who do are assured of both a profit and a livelihood on good ground.

Everything else Berry says about agriculture elaborates this basic paradigm. He argues that we need to have wilderness available to visit to
preserve a memory of what the world was like before human beings
began to alter its contours. Wilderness will thus be available to serve as
a model for what he calls the "kindly use" of land and natural resources
that is required for their long-term usefulness to us. This is a use that
goes along with the tendency of a landscape to grow the kinds of plants
and animals that are best suited for it. Kindly use depends on an intimate knowledge of its "kinds, climates, conditions, declivities, aspects
and histories" (p. 31); and this in turn requires farmers who are allowed
to live on their land and use it according to their ability to balance both
the land's needs against the consumer's needs. It is the farmer, not the
food processor nor the consumer nor the banker nor the agricultural

engineer nor a government bureaucrat who must determine what that use should be. Basically we must "keep the source of food independent of any but agricultural means..." (p. 37).

"Bigness" is the ally of specialization in the modern world's war against good, wholesome farming. Bigness has no tolerance for the small details of a landscape and the nuance of particularized stewardship. Mass production requires mass-produced machinery and uniform cultivation of great expanses of land. Berry hates bigness for its shoddy products and for its inclination toward consolidations of power and money:

> As a social or economic goal, bigness is totalitarian; it establishes an inevitable tendency toward the *one* that will be the biggest of all. Many who got big to stay in [farming] are now being driven out by those who got bigger. The aim of bigness implies not one aim that is not socially and culturally destructive. And this community-killing agriculture, with its monomania of bigness, is not primarily the work of farmers, though it has burgeoned on their weaknesses. It is the work of the institutions of agriculture: the university experts, the bureaucrats, and the "agribusinessmen" who have promoted so-called efficiency at the expense of community (and of real efficiency), and quantity at the expense of quality (p. 37).

For Berry the word "agribusiness" tells the tale. It replaces the traditional word "culture" with "business," as if to say that what used to be a good life for farmers now has to become a paying proposition. This is not so much a lie as a "radical simplification" as he calls it, because good farmers have always worked hard to make money, but only as part of a complex act of working hard for the sake of a good, wholesome place to live on: "Farming, the *best* farming, is a task that calls for this sort of complexity, both in the character of the farmer and in his culture. To simplify either one is to destroy it" (p. 45). Berry's model for the best agriculture is based on the ecological principle that "you can't do one thing." Everything alive and natural depends, ultimately, on everything else alive and natural. Every system meshes with every other system: "We can have agriculture only within nature, and culture only within agriculture. At certain critical points these systems have to conform with one another or destroy one another" (p. 47).

The Prosperous Farm

To counter Berry's arguments point for point we could quote from the same speeches of Earl Butz and other advocates of agribusiness that Berry cites throughout his book. Their central theme is that technology increases efficiency, efficiency increases productivity (widens the proportion between low input and high output), and productivity increases profits on the farm. Although they acknowledge that with modern technology fewer farmers are needed to grow our country's food, they think that this is by and large a good thing. The few farmers who remain have an easier life more in tune with the lives of their fellow citizens in the suburbs and cities. Those who no longer live on the farm are more productively employed elsewhere. The point of sharpest disagreement between these arguments and Berry's would be over his statement that we must "keep the source of food independent of any but agricultural means." Their counterargument would be that we should "keep the source of food independent of any but economic means." That is to say, agriculture is not a privileged activity in our society. It must submit to market mechanisms and in the process subordinate both the life of the farmer and the life of the farm to the needs of the larger life of the community outside the farm's fences. In the last analysis, any kind of culture, agri-culture included, must change with the times and with the demands of markets. No matter how nice it looks from the road, the only farm that deserves to survive is the farm that stays financially solvent. These kinds of arguments will chase each other around the tree until the cows come home.

It might be more interesting, therefore, to counter Berry's arguments with those of Theodore W. Schultz, who won the Nobel Prize in Economics in 1979, shortly after Berry's book appeared. Schultz is a very subtle agricultural economist who like Berry looks behind current policies to the fundamental principles according to which farmers or any other citizens live their lives. But unlike Berry, he believes the economic principle is paramount, on the farm as elsewhere, and always supercedes the cultural principle eventually. Pitting their views against each other allows us to see them as more than a passing political debate; they are two enduring contraries of human life.

What makes Schultz such an interesting adversary of Berry is that Schultz also believes that what he calls "traditional farms" are very efficient users of resources and producers of goods. Their resources might be few and their profits small, but they are well used and honestly won. Traditional farmers are basically very shrewd, with an eye out to any advantage they can gain, however small. All their calculations are "made with a fine regard for marginal costs and returns" (p. 39). What distinguishes their farms is that they are very stable institutions that can remain on the same soil and partake of the same sustaining culture for generation after generation: "While the communities in this class differ appreciably one from another in the quantity of factors they possess, in what they grow, in the arts of cultivation, and culturally, they have one fundamental attribute in common: they have for years not experienced any significant alterations in the state of the arts" (p. 37). Schultz uses traditional farms in India and Guatemala for models, which, although they are not precisely like the traditional American farms Berry devotes most of his attention to, are close enough to them in being both efficient and stable. Berry also uses traditional third-world and Amish farms as possible models for his ideal of farming and as representative of what he considers the best aspect of farming anywhere, in America as well as in the third world. Therefore what Schultz says of traditional farming throughout the world applies as well to what Berry says of wholesome farming in America.

Much of what Schultz says about traditional farming argues against writers who question its efficiency (and thus once again, squares with Berry's arguments). Its farmers "are remarkably efficient in allocating the factors at their disposal in current productions" (p. 43). They are also very smart at what they do. They continually refine what Schultz calls the "lore," or "empirical wisdom" they inherit and thereby over time manage to adapt slowly and precisely to every environmental and social change that comes their way. Given the tools and seeds and soils with which they begin, Schultz insists, and the lore they have developed, it would be virtually impossible for these farmers to do any better than what they do.

Schultz and Berry part ways, however, where Schultz desires to "transform traditional agriculture" in order to make farms instruments for the production of wealth. He sets about discovering the best means to bring the transformation about. Essentially it means changing the mind and culture of the farmer through wholesale reeducation. Nothing less will do: "rapid sustained growth rests heavily on particular investments in farm people related to the new skills and new knowledge that farm people must acquire to succeed at the game of growth from agriculture" (p. 177). His reasons are unsentimental. The human population continues to grow in the third world and elsewhere, while the amount of land available to feed this population remains the same: "good farm land is no longer around for the taking" (p. 179). Thus the minds of farmers are the only truly untapped resources upon which to draw to increase productivity as it needs to be increased.

One obstacle to educating farmers sufficiently is the prejudice that farmers are inherently stubborn and stupid. Another is the political systems in many countries that allow large landholdings to concentrate in a few rich families who have no interest in educating the campesinos on their lands. But these obstacles aside, education is the key to any gains in agricultural productivity. Education even has the power to compound interest by increasing the life expectancy of the very farmers whom it enables to learn new ways of doing things.

Where Schultz differs from many other advocates of educating farmers is in insisting that farmers need more than technical training. They need an understanding of how the world works that only a broadly based education can bring. They need to understand mathematics and not just maintenance, economics and not just bartering. This is just the sort of education, of course, that will break down the major distinctions between farm workers and urban workers and will destroy in the process the "culture" of agriculture that Berry continues to insist we should prize. Schultz is clear on this point: "The prevailing cultural values as a rule not only exclude the scientific and technological component of modern culture but they debase this important component in what students are taught. Farm people even more than many workers in nonfarm jobs must acquire skills and knowledge drawn from science if they are to be effective in using modern agricultural factors of production.... Thrift and work are not enough" (pp. 203; 205). Unfortunately, Schultz does not appear to consider any of the humanities an important component in the broad education of a farmer, unless of course he might consider his own discipline, economics, among the human sciences.

Conclusion

Bringing the arguments of Berry and Schultz together raises certain fundamental questions neither addresses directly. Is Berry right in thinking that agriculture can remain set apart from the demands of the economic community at large—like a sort of monastery on the hill? Can economics honor ecological constraints as he seems to think it should? Is Schultz right in thinking that "culture" opposes technology and economy and has to be expunged in order to make room in the farmers' minds for know-how?

Another related issue is that the new, more efficient farm might produce great surpluses of cheap food all right, but at the same time could be mining the soil of its tilth and fertility faster than it can be replaced by the techniques of modern agricultural research. Might there be some iron law, according to which all farming tends toward low productivity but sustainability; and that any farm that increases its productivity gradually eats up its own biological basis? Schultz is typical of ag-economists in deciding not to reckon the long-term costs of human anguish and environmental degradation that often result when larger, more economically efficient farms absorb smaller farms. We should not fault him for this. The task of the economist is to give us a "pure" analysis of what happens when the economy changes.

Finally, is there really a debate between Berry and Schultz at all? Perhaps the debate between culture and economics in agriculture as elsewhere resembles the debate between creationism and evolution. The assumptions behind them are fundamentally different and therefore do not admit of any honest comparisons. It is the difference between faith and reason. This is not to say that Berry's argument has to do with faith, and Schultz's has to do with reason. Both are based on faith. Berry believes in the value of the rural life. The countryside is a good place to bring up children. There are riches to be had in knowing a creek, in hunting, in caring for animals, in returning a worn-out slope to good tilth that are the equals of those to be found in libraries, shops, and banks. Let us not be quick to assume that farming is the play of children that most people should be allowed to outgrow. Much of farming's wisdom is perennial. Schultz believes in the value of cities, in a highly specialized, technological environment with its hospitals, laboratories, and cinemas. He sees that even rural life in developing countries is coming to resemble urban life with its crowding and with the demands of more mouths for more food. The world's growing population deserves enough food to eat, which only truly economically viable farms can supply. Farming wisdom has to become more sophisticated to keep up with the new demands. Both Berry and Schultz marshal every reasonable idea they can in support of their very different visions of the good life. In is *within* the thought of each of them that we find the gaps that do not close—the very gaps in logic his opponent will expose.

By reading these two essays we truly grasp one of the contraries of human experience. Perhaps any "culture" a farmer manages to enjoy in the face of the hostile forces of nature and society constitutes a rare triumph of luck, whether he or she farms a small, traditional rice paddy or a huge, verticalized hog operation.

Readings

Berry, Wendell. "The Ecological Crisis as a Crisis of Character"; "The Ecological Crisis as a Crisis of Agriculture"; "The Agricultural Crisis as a Crisis of Culture." Chaps. 2, 3, and 4 in *The Unsettling of America.* New York: Avon, 1978.

Schultz, Theodore W. "The Allocative Efficiency of Traditional Agriculture"; "Investing in Farm People." Chaps. 3 and 12 in *Transforming Traditional Agriculture.* Chicago: University of Chicago Press, 1983.

Wojcik, Jan. "The American Wisdom Literature of Farming." *Agriculture and Human Values* 1, no. 4 (Fall 1984): 26–38.

Chapter 3

The Policies of Agriculture

The Cases

No Verticalization!

A legislature in a traditional farm state is considering a new farm bill. A controversial part of the new bill calls for repeal of the state law prohibiting single companies from performing more than two of the traditional seven functions in the whole process of bringing food to market. These functions have been defined as doing fundamental research in any aspect of agricultural activity; supplying basic materials to farmers such as feed and machinery, chemicals, and maintenance; growing a product; transporting and storing products; processing basic products into specialty foods; marketing farm products; and finally selling products on the open market. The idea behind the law had been to prohibit the "verticalization" of any particular agriculture industry, that is, the control of all the aspects of the growing of a product such as corn, chickens, hogs, or dairy products. The reasoning was that such a prohibition would protect individual operators of any of the functions from competition with large corporations, and instead would encourage competition among many operators. This diffused competition in turn would be good for both operators and consumers; it would keep the price down for consumers and it would enforce efficiencies at every level in the system, efficiencies that would encourage innovative entrepreneurs and farmers to take risks to stay in business.

This law had a history of strong support throughout the state, especially among farmers, because it enabled them to diversify their farms, to try growing a number of different products very carefully in order to be able to compete with other farmers in the quality of their products. Conservationists liked the law because it tended to keep farm size small and in the hands of family operators. Conservationists believed these farmers were more inclined to take care of their land while they competed in the market. The agribusiness community liked it out of habit more than anything else. The law had established the only way of doing agribusiness in the state, and most of them were used to it, and by definition, those still in business were competitive.

But now Lorraine Haworth, a legislator from a large city district representing the major financial and business interests in the state, argues that such a law represents nostalgia for the past rather than an awareness of the realities of the modern world. She argues that it does so in two ways. Firstly, nowadays agricultural technology has advanced to the point where one company can efficiently plan and initiate all the aspects of the growing and marketing of a product with the single goal in mind of producing the largest surplus at the lowest price and the largest profit. Studies have shown that central planning of this sort is now more efficient than the market in this regard. Secondly, she argues, this has already proved to be the case. Neighboring states that have allowed verticalization of the chicken market are producing chicken that is much cheaper than local chicken and shipping it into the state under a well-advertised, popular brand name that is outselling local chicken four to one. Various consumer groups lobby in support of Haworth's bill to allow verticalization in the state.

Initially most of the legislators representing farm districts as well as the lobbyists for farm organizations are opposed. Jill Goodfellow, a spokesperson for the lobbyists, argues that the family farm has always been sacred in the state, and this bill would force out a large number of farmers and operators who had developed a reputation for the state's chicken as a superior chicken. Goodfellow proposes instead a new marketing campaign run by the state that would emphasize the wholesomeness and tastiness of local chicken and thus try to make up in higher prices for a better product what might be lost in volume to the neighboring states.

The legislator from the financial district nonetheless wins sufficient votes for the passage of her bill. One argument she uses is political. Haworth argues that forbidding efficient verticalization amounts to a government subsidy of the less efficient, smaller farmers who are thereby protected. She uses this argument at a time when the federal government is decrying subsidies of every kind. Another argument is economic. She presents a study that shows consumers almost universally buy chicken by price rather than by quality. She concludes that verticalization of the chicken industry therefore is the only way that local chicken can hope to compete in a regional market, indeed, even within the confines of her own state.

Her bill passes and the number of chicken-raising operations in the state are reduced to a tenth of the original number within four years.

Taking a Chance

Sandy Gambino, a farmer in a traditional farm state, has worked a farm his whole life. He inherited the farm from his parents, who had worked it during their lifetime. It has been a highly diversified farm with a fairly large broiler and egg operation as well as grain crops and a herd of beef cattle. Over the years the farm has changed its emphasis as times have changed and markets have fluctuated. In good years for grain sales, the chicken flock and cattle herd were reduced by selling stock for meat; in bad years for grain, much of it was fed to the livestock, whose numbers were bred up. In recent years, when grain sales have been consistently weak and there has been a national trend among consumers to prefer white over red meat, the farm produced more broilers than cattle.

When a new law is passed in the state legislature allowing for the verticalization of the chicken market, the farmer-owner decides the only way he can hope to compete in the new market is to get out of the chicken business altogether and to turn his operation into a grain and beef operation. But because Gambino has much of his capital tied up in his chicken barns, which no longer have any business value, he decides he needs to take out a loan to refurbish these barns for feed storage and cattle shelter. He finds that his local banks turn him down, however, because their loan officers estimate that he will never been able to show a profit in grain and feed because, even with expansion, his volume will be too low even to service the loan. Gambino applies to the Farmers Home Administration, which often loans money to farmers who cannot get loans elsewhere, and is again turned down. The policy of the current administration in Washington is that no loans should be made that would subsidize uncompetitive farms.

Gambino decides that it is not only time to get out of the chicken business, but to get out of farming altogether. He is in his late fifties and decides that the time has come for his children to take over the farm and divide it up as they will, and for him to retire rather than change the nature of the farm himself. His four children have a family meeting and decide that an unmarried son and daughter will run the operation. They discuss whether the other two, one a dentist, the other a physician, will sell their interests to their siblings or invest in changing the way the farm does business. If they do change the farm, they could change it to either a corn and cattle operation exclusively or apply for a franchise in a large-scale broiler operation that an agribusiness is offering. In this case, the company offers to supply small chicks for the farm to raise to broiler size in six weeks when company trucks will pick them up and take them to their processing plants.

The Gambino children decide first of all to change the farm because its old structure is no longer viable in the current market; and second of all to invest in the huge building that operating a broiler franchise requires. The farm becomes a part of a verticalized chicken-product corporation, and for the first three years of its operation returns 25 percent of its original investment a year in income to the four operator-owners.

In the fourth year the chicken corporation switches to its own buildings for raising broilers and no longer farms out that aspect of its enterprise. The Gambino children's farm cannot make the transition to another mode of agricultural operation, and they declare the farm bankrupt.

Moving in; Moving out

A community of Amish finds that it can no longer find farms to buy in its area to satisfy the needs of its younger families for land. But young Amish families now find that there are newly bankrupt farms coming on the market in an adjacent state. One family, the Benjons, buys a farm from a bank that had originally been owned by a family corporation but had to declare itself bankrupt and was foreclosed on by the bank when its four owners had decided not to salvage their interests in the farm. Several other Amish families help the Benjons to tear down the large chicken barn on the farm and to use its lumber for constructing a number of much smaller outbuildings. The new farm eventually raises vegetables, feed, hogs, cattle, chickens, and geese primarily for its own use and berries for market. Members from several nearby Amish families, who also recently moved to the area, help the new owners to build a small processing shed on the farm where they work together to produce jams and preserves from the farm's berries and from fruit grown on other farms to sell in a local city's farmer's market. In this way the Benjons earn money to pay the local land taxes.

The area into which the Amish move undergoes many changes. A number of schools close as there are fewer families with children who will attend school past the tenth grade, when most of the Amish children leave school. Taxes fall, services decline, many old-time residents move away, and soon the area contains a few large corporate farms selling to a national market, and many small self-contained Amish farms.

Eventually the Amish farms fail too after new state legislation requires all preserves to be marketed in jars that have been sterilized at a very high temperature. The Amish do not allow themselves to use any machinery of the sort that would raise the temperature of their jars to the required levels. The highest temperature they can reach is the 212 degrees of water boiling on a woodstove. They cannot market their jellies; cut off from any cash crop, they cannot pay their mortgages and property taxes and need to move out of the area and the state.

The Issues

Change

Change in agriculture is inevitable, and much of it is inadvertent. It results from forces working on the farm that no human being can control. The climate changes. The climate is bad for wheat growing one year, and the demand for bread continues to grow with the population. It is more than likely that more wheat will be planted the following year, and it will probably be a variety particularly resistant to dry weather or wet, if last year's bad climate appears to be part of a trend; et cetera. Farmers can hardly help changing too. Theodore Schultz's work shows that even highly traditional peasant farmers continually refine and improve their farming techniques, both to adapt to changes in climate and land and to remain competitive with the other farmers working around them. At the same time, farmers can hardly avoid the consequences of many decisions made by their fellow citizens living away from the farm. Wealthy urbanites might decide they want a country house as a refuge from intolerable living conditions in the city. They thereby drive up land prices in the countryside beyond those that farmers can afford. Suburbanites develop a taste for mass-market processed foods. They thereby unwittingly drive out of business local farmers geared up to supply vegetables to local restaurants at premium prices; and so on. One could generate endless examples of how the natural and social worlds inadvertently affect the farm. (See chart 3.1.)

CHART 3.1   Changes in Agriculture (in order of their susceptibility to human control)

Undirected

From climate

From pests

From erosion

From changing food tastes

From changes in general education

From changes in social conditions

Directed

From providing farmers with technical education

From changes in particular laws governing farming practice

From pursuing consistent policies about farming practice

Recognizing the inevitability of most of the changes in farming does not stop some people from wondering whether farming might not be changed deliberately for the better. Obviously, right now it is useless to think about changing climate (although, of course, the greenhouse effect or acid rain or the ozone hole might be doing it for us, willy nilly). Some might try to change farming to change the land. They might work to enact new laws to encourage conservation or to restrict the use of abrasive agricultural practices. Politicians might zone land strictly for agricultural use, and so on. But these efforts, like all deliberate efforts to change farming, are directed at changing people first.

The least obtrusive way to change people in order to change farming is to educate them about such things as nutrition, conservation, and the new techniques continually being developed. This way people can make educated decisions on their own about how they should farm, buy, and prepare food, and the responsibility to change becomes diffused throughout the entire culture. Consequently education changes agriculture indirectly by diffusing decision making widely throughout the culture and profoundly by rooting the change in the attitudes and expertise of farmers and consumers alike. This was the wisdom behind the Morrill Act that established the land-grant university system in this country and led to the establishment of the extension service. It is the wisdom behind Schultz's call to educate third-world farmers broadly, not just in expertise but in a larger appreciation of the world into which their agriculture fits.

The most intrusive way to change people in order to change farming is by political fiat. It does not have to be forceful; it can simply allow certain kinds of farming to be practiced and thus allow the economic marketplace to choose among them in the process of making choices about their different products. Laws about marketing agricultural products can also affect farming. If a law insists that a company list on a box of cereal all the ingredients including its chemical additives, consumers might choose to purchase the cereal with the fewest additives and thereby encourage the kind of farming that would produce the kind of food that needed the fewest. But although the political process might have more immediate consequences on farming than education, the kinds of change both effect are similar in that they are most often indirect. It is clear why this is so with education; but it is often so with politics because it is hard to predict ahead of time just what the consequences of a political decision will be. Allowing a certain kind of farming such as the verticalized chicken farming described on page 35 does not mean that consumers will buy its products and insure its success and survival.

Therefore a policy is more forceful than a law, in the sense that a person in pursuit of a policy will keep tinkering with laws or amendments that will bring about the goal the policy is intended to achieve. Law is just a tool of policy; pursuing a policy means making continual adjustments to the law; and it is controversy over political policy in regards to agriculture that is the particular topic here.

The controversy over agricultural policies resembles the controversy about agricultural principles discussed in chapter 2. A policy, like a principle, pursues long-term, fundamental effects. Theodore Schultz, for example, wants an education for farmers that will change their whole worldview in order to make it more compatible with the use and expertise of modern technology in the service of modern markets. The eighteenth-century economist Adam Smith recognized that people are self-interested and greedy by nature and wanted to harness the greed by socially controlled competition. Schultz wants to create self-interest and greed where it did not exist before—in the traditional farming community. He wants to put policies in place to make farmers sophisticates, capitalists, consumers, and cash farmers all together.

Wendell Berry, on the other hand, wants an education for both farmers and the public that would instruct them in the constant, basic values required if farming, farmers, and food are to remain wholesome. Berry is, in his own way, a disciple of Mahatma Gandhi, whose approach to the problem of limited resources in India was to suppress basic human drives. Humankind should learn to live with less and it should put less emphasis on the material things of life. But Berry argues that a highly sophisticated, materialistic culture like modern America, with enormous resources already at hand, should give many of them up. Berry wants policies that would encourage citizens and farmers voluntarily to become simpler and satisfied with fewer but wholesome goods and foods.

But the controversy over policy soon becomes very different from the controversy over principles and thus deserves to be considered in its own right. One difference is that policy is practical-minded. Someone pursuing a policy thinks in terms of passing or thwarting this or that legislative bill and often argues in terms of very practical, easily measurable short-term effects. Thus the controversy over policy is more specific and pointed than the controversy over principles.

Another difference is that policy is aggressive. The debates over principle are often lofty and detached, as in the writing of Theodore Schultz. When these debates become heated, as in the rhetoric of Wendell Berry directed against Earl Butz, the tone becomes that of a prophet's general moral outrage. Politicians, in contrast, have specific constituencies and adversaries in mind. Their rhetoric will appeal to the self-interest of supporters as if they were really conscientious citizens with the best interests of the entire community in mind. They will dismiss the self-interests of the opposition as if they were really no more than selfish people with only special interests in mind.

Thus, the polemical tone of political rhetoric about agriculture bears particular watching, lest it cloud the issues. Such authors often write as if they think the grass is greener on the other side of the policy fence. The *other* policy always seems to be enjoying more support; a better policy needs a hearing badly, and now, before it is too late. There is a sense of urgency in this writing, even a sense of alarm. No one ever making a political case likes to admit that his or her policy is doing as well as the opposition seems to think it is. Therefore the controversy over policy often passes over facts or forces facts to fit an agenda and does not have any real interest in assessing what precisely is being done in agriculture today and what would be the best policy to pursue in trying to change what is being done for the better. That is a task for more dispassionate readers and thinkers.

This chapter arranges seven essays into an extended debate about policy. On the free-market side are essays by the American journalist Gregg Easterbrook, the American agricultural economist E. Wesley F. Peterson, the British researcher Gwyn James, and a committee of agricultural professors at Iowa State University. On the protectionist side are essays by the social scientist Kenneth Dahlberg, the historian George Borgstrom, and a pair of agricultural professors at Oklahoma State University. We are still dealing with the basic principles in conflict that we encountered in the previous unit. The difference now is that each essay here is more specific. The broad, almost philosophical generalities of the previous unit are now linked to actual policies that might be voted on in Congress.

The Review

The Free-Market Side

We begin with a journalist's view of the current farm policy published in a recent issue of the *Atlantic.* It is a timely analysis of the hodge-podge of present-day farm policy, because as it is published in the summer of 1985, Congress is in the process of developing a new farm bill. Gregg Easterbrook basically agrees with Schultz that America has too many farmers to have an efficient agriculture given the present state of agricultural technology. We could do with far fewer farmers, he says and thus openly breaks with Wendell Berry's view that when America loses a good farmer, she loses a part of her soul.

In fact, in good journalistic fashion, Easterbrook sets out to demolish our most cherished myths about American farming that cannot withstand the facts as he sees them. He says we need to forget our images of the red barn in the sunset, the wholesome pitcher of milk on the refreshment table of the Grange Hall dance. Forget the impressions we might be getting from the sentimental media that down on the farm things are less worldly and more wholesome than they are in the cities. Farmers farm to make money, just like the rest of us, he says, and they are not doing badly. "Farm families are not poor . . ."—the average income for farm families in a recent year was $21,907, only a few thousand dollars under the overall national average. "Farmers are not being driven from the land . . ."—in a recent year remarkable for its number of news stories about foreclosure, the Farmers Home Administration actually foreclosed on only forty-two farms nationwide. Although debt problems are real—"farming is not a disastrous investment"—it is still returning a steady 2.5 percent on investment year in and year out. A farmer or investor is not going to get rich on such a return; but he or she is not going to go bankrupt. On another point, he says most farming is not done by agribusiness concerns, despite the frequent alarms in the press about the squeezing out of the family farm. Easterbrook's view of farming is that it is a relatively stable money-making institution, kept proper and prosperous by a large number of farmers and their financial friends. The vague sense many of us have that the family farm and the traditional nature of agriculture is at peril is the consequence of too many pundits decrying as trends what is really only the normal flux of good times and bad.

Easterbrook's facts and figures are interesting, and we need all the help we can get to keep the facts in mind and our eyes clear when considering the emotional issues of farming. Note, however, that to strengthen his case, he chooses his facts carefully. He talks about average salaries, not median salaries. The median salary for farmers might be quite a bit lower than an average that is skewed by the potentially enormous profits of an agribusiness. He calls a perennial 2.5 percent return on investment "steady," which it surely is. He does not raise the question whether such a small return would satisfy most investors. We also need to keep in mind how he is using those facts. Easterbrook uses them to reinforce his belief in the principle that farming is primarily a money-making proposition. We should consider marketing the product as important as growing it—its simply logical extension. This lays the groundwork for the real purpose of his article which is to convince the reader that all the federal agriculture subsidy programs ought to be done away with. They are constraints on the free trade that is the only nourishment a healthy agriculture needs.

Coming out against subsidies is a popular position that almost everyone connected with agriculture takes at one point or another. Americans feel uncomfortable with any kind of welfare system. The subsidy system comes into play when farmers find that the price they are receiving for their product is lower than the price they need to pay for their expenses and make a profit. This happens because many of their fellow farmers produce more than the market can absorb. With most products that would mean that demand would fall and price would fall, and those producers who could not produce more efficiently—by lowering their production cost—would go out of business. This would leave fewer producers producing, which would lead to a lower supply, thus a higher price, until ideally a balance was reached between supply and demand, expense and profit, price and satisfaction. But a subsidy program prevents the free-market system from working. The government pays farmers more than the market price would bring. This way the higher price will give many farmers the price they need to be able to make a profit after paying their expenses.

This kind of subsidy is called a deficiency payment and is probably the easiest to understand and to implement. The government sends a farmer a check to make up the difference between the actual market price and a target price set by Congress ahead of time. More complicated are Commodity Credit Corporation loans whereby the government loans a farmer a sum calculated to be a fair price for his or her crop. The farmer puts up the crop as collateral. For example, in one year a farmer expects to harvest 150 bushels of corn per acre on a thousand acres of land, or 150,000 bushels. He has done so in the past. In the last few years weather patterns have been consistent and good for growing. Government officials calculate that the price for a bushel of corn at harvest in the coming year is likely to be $3.00 a bushel. They lend the farmer $450,000.00 to pay salaries or himself, to buy more land, or more equipment, or perhaps to pay off an old loan at a high interest rate. If the price turns out to be higher than $3.00 a bushel and the farmer indeed harvests as much as he had hoped, the farmer sells the corn on the open market, pays off the loan, and keeps the excess. If the price is lower or his harvest is not so big, the farmer gives the crop to the government and keeps the loans. This way the farmer always earns at least as much as a target price, and sometimes more. Often the government attaches strings to the payments or the loans, requiring that farmers participating in the CCC programs leave idle a certain percentage of their arable land, in this way cutting back on the potential surplus of a given product. Our farmer might have only been lent money against his harvest on five hundred acres of his land.

Other subsidies control the free market of agricultural goods in more subtle ways. The Farmers Home Administration is the lender of last resort to farmers who have been turned down for loans elsewhere to buy feed and equipment. Another federal agency, the Small Business Administration, grants emergency loans to farmers when natural disasters destroy their crops. The government also subsidizes the efforts of farmers to conserve their water and land, and it has funded the great water resource projects in the Far West. It supports agricultural research in the land-grant university system and the dissemination of expertise through the Extension Service. Finally the government sells agricultural products through the food stamp program domestically and through the Food for Peace program abroad.

Easterbrook comments that our government's policy of subsidizing agriculture survives despite almost universal dislike of any policy that protects inefficient industries, be they producers of pig iron or pigs. Farmers dislike it even when they are hooked on one of its programs because they are independent types worried about the stigma of anything that smacks of welfare. The subsidy system makes every farmer miserable to more or less the same degree. Everyone makes an adequate but only a small guaranteed profit; subsidies discourage innovation and risk taking and the exhilaration they bring. Of course, Easterbrook passes over any discussion of innovations individual farmers might finance with their subsidy checks in an effort to become more independent. Instead, he emphasizes the well-documented fact that hard-working people hate admitting they depend on welfare, and that is what the subsidy system is often accused of becoming.

Yet year after year the policy of subsidizing agriculture survives every attack made against it, for reasons that Easterbrook finds very interesting. They reveal our national character. One obvious reason has already been implied. Opposition to the whole policy is divided, and thus conquered, by each farmer who benefits from a particular program. But the support of individual farmers is not enough; they are now only a small percentage of the nation's voting population. The policy survives, Easterbrook says, because the general population has a warm spot in its heart for farmers, again for reasons already given, and thus supports in principle and in the votes of even urban congressmen the subsidy policy that actually does permit the survival of many more farmers than a free-market policy would. Farmers are happier on the farm (it is thought) and perhaps better off staying out of the cities looking for work—which is where many of them would go if forced off the farm.

Why we like farmers staying on the farm is a little more difficult to say. Perhaps it is because many of us grew up on farms or have parents who did, and even those of us who did not have positive images in our heads of the noble farm landscape as the sun sets in the autumn on its pumpkins in the field. We subsidize farming with the same sentiment with which we purchase albums for our family snapshots. Another motive might be that human beings are naturally suspicious of new technology. A major reason why farmers find it easier to produce ever greater surpluses is that their new tractors and seeds as well as new innoculations for animals and new management and computer systems focus all of a farm's activities around a core of efficiency. This in the abstract is good, but in the concrete it means that this nation now requires far fewer farmers to produce the food it needs to feed itself and sell abroad. Thus these new tools threaten to crumble what has always been the bedrock of rural life—the community each farmer shared with his fellow farmers. It still seems more natural to love thy neighbor than to love thy tractor; and a subsidy policy is one way to handicap the tractor in its competition with the neighbor across the fence. When all is said and done, subsidies brake technological progress if they make it less likely that a farmer will buy a new tool to make his operation more efficient—less likely because the farmer is making a good enough living with his present production plus his subsidy payment. This, anyway, is what Easterbrook believes. There is nothing, of course, to prevent a farmer from plowing some part of a subsidy back into the operation in the form of a new plow or tilling plan.

Easterbrook is clear-eyed enough to recognize that farmers are also naturally competitive, that they often cast covetous eyes across their fences on their neighbor's field, and that many of them deep down believe that given free hands, they would use the new tools and the new opportunities to out-produce any other farmer farming around them. Thus we could add to our list of the reasons why farmers dislike the subsidy policy the simple facts of greed. But finally the farmers' pride in their traditional instincts for community is greater than their natural lust for success, and so whatever their grumbles in public, they and their urban neighbors support the policy that warms the heart the most.

Easterbrook would have us be cooler. The tactic of his argument is bold; he tries to disarm the argument of a Wendell Berry by facing it head-on. He features two attractive, young, and struggling farmers, Dennis and Patricia Eddy, who live and work on a farm in Stuart, Iowa. He depicts these two people as the salt of the earth; at the same time he contrasts their noble, struggling family farm with a highly successful, diversified, and industrial-scale farm in California run by anonymous M.B.A.'s for whom a day on the farm is like a day at the office. Yet at the end of the essay he states his belief that we could do without such salt on the earth quite nicely. The California farm is the farm of the future and the present, while "the Eddys are young, responsible, bright, and eager to work. If they can't land on their feet, who can?" Easterbrook takes the Olympian view that sometimes progress is cruel to individuals while working for the benefit of the group—much like it is when a family loses its homestead to highway construction. "Eventually Congress will have to face the fact that there are too many farmers. . . . There can still be family farms. It's just that not every person who wants a farm can have one, no matter how fervently we might wish he could" (p. 78).

E. Wesley F. Peterson, a professor of agricultural economics, puts the matter even more coolly. Like Easterbrook he believes there is something like an invisible hand directing the structural changes in agriculture. But he believes there is nothing anyone can do to stay or reinforce that hand significantly. Writing in the journal *Agriculture and Human Values* about structural change he says:

> I do not wish to deny that some changes could be contemplated to improve the chances of survival of mid-sized farms [like the Eddys']. For example, public research in agriculture might be directed toward developing technologies or methods biased in favor of mid-sized farms. Further research on the causes of structural change might result in identification of other areas where changes could be made. However, I suspect that these efforts would not be sufficient to stop structural change and it is not clear that the social benefits of such action would outweigh the costs (p. 13).

Gwyn James, a British economist and author of *Agricultural Policy in Wealthy Countries,* is even more stern than Peterson. He states flatly that agricultural policy should *not* be made for the good of any kind of farmer: "Farming, like any other form of production, is not an end in itself; the purpose of farming is not to provide the farmer with a 'proper remuneration' or a 'fair standard of living,' but rather to satisfy the wants of the consumer" (p. 308). Furthermore, social and ethical considerations about fairness to farmers do not withstand "valid economic assessments" (p. 307). Although he admits a "guaranteed price system may be an effective, and perhaps indispensable means of increasing production in times of acute food shortages... [it] is less easily defended (indeed can be openly attacked) in times when consumers can freely satisfy their requirements and preferences by considering the relative merits of a variety of competitive markets" (p. 313). Finally, he equates a farm with a firm in speaking of the iron laws of economic necessity: "The elimination of the inefficient firm is a continuous and inevitable feature of a declining industry, but in a growth industry it is limited to those marginal firms who are unable to compete effectively with their rivals" (p. 314). In either case, he argues, let the invisible hand of the market do the necessary pruning.

But James goes beyond Easterbrook with four arguments in favor of aggressive government intervention. First, the government should force marginally inefficient farmers out of farming by seeing to it that the "salvage price" of staying on becomes unbearably high. This is a mean phrase. It refers to the price a farmer is willing to pay to hold on to a beleaguered farm in hopes that a brighter day will dawn. Such tenacity has become a distinct liability nowadays. Inefficient, stubborn, "traditionally minded" farmers, he says, must be removed from the landscape. Second, the government should discourage anyone from deciding to start farming if it has been determined that the supply is adequate to the need. Third, the government should monitor the marketing of agricultural products closely to make sure that costs, profit, and price remain in balance. These are virtually arguments for social Darwinism, according to which only the economically fittest farm must be allowed to survive.

However, with his fourth argument James shows that he is not heartless. He thinks the same government that forces inefficient farmers to leave farming should take pains to smooth their way. There should be social programs to help with the transition, informing farmers of the kinds of work they could do elsewhere with the skills they obtained on the farm, sometimes even helping farmers with retraining programs. Basically he believes that it should be government policy to insist upon the play of the free market but also to ease the pain of the inevitable losers in the game.

Although the Americans Earl O. Heady and his fellow faculty members at Iowa State University advance the same ideas as the Britisher James, they claim the government can make agriculture more efficient simply by leaving it alone. In their book *The Roots of the Farm Problem*, they say plenty of American farm kids are already preparing to bail out. They are going to school to learn how to do other things with their minds and hands than farm. Those who want to remain on the farm are no fools either. They understand the need to upgrade their production and marketing skills and are getting the education and the tools they need. The agriculture extension services are already on the job disseminating the necessary information for anyone that will listen. Finally, the "growing competition and commercialization of agriculture under existing and prospective technology and input prices" proceeds apace (p. 197). As it does, "the sociological characteristics of the 'farm way of life' undoubtedly will disappear and the nostalgia of farm fundamentalists will become less intense" (p. 212). (See chart 3.2.)

CHART 3.2   Agricultural Policies

Free-market (or cheap food) policies

Forget sentimental, old-fashioned images of farming

Treat farming as a prosperous industry

Withdraw protective subsidies

Force inefficient farmers off the land

Retrain marginal, uncompetitive farmers for other work

Protectionist (or good farm) policies

Honor traditional farming values

Treat farming as a special industry

Measure short-term economic needs against long-term sustainability

Protect farming environmentally

Preserve rural culture

The Protectionist Side

Kenneth Dahlberg's *Beyond the Green Revolution* was published six years before Gregg Easterbrook's essay, but we can still read them together and read the older one as if it were a response to the newer one, because the questions each raises about policy are at root perennial questions. We judge them on their articulation of venerable positions, not on their timely relevance to the passage of a particular farm bill.

Dahlberg's basic argument is that industry and agriculture are not the same. Industry finds and mines raw materials and turns them into finished products with no real regard for the original raw material, only for the eventual product. Agriculture, on the other hand, manages an exchange between living things in such a way that it must nourish both parties to the exchange. The farmer feeds more than the consumer; the farmer feeds animals and crops and in the process of feeding them, feeds the land with fertilizer and the conservation processes that preserve the soil that feeds the rest of the food chain. Because of this fundamental difference, it is not accurate to use the economic standards suitable for industry to measure agriculture. In fact, it is dangerous: "By depending upon inappropriate economic models for our understanding of agriculture, we risk blinding ourselves to the many adaptive changes that are possible, and we encourage the even greater danger that our increasingly simplified and centralized agricultural system will collapse" (p. 165).

Dahlberg's first chapter is a bold attempt to explain why many economists do try to measure agriculture as if it were an industry. Basically they do so because they are used to measuring every human enterprise as if it were an industry; and the results are crisp and graphable. The results *seem* to be genuine because they seem to offer an over-view or an inner-view of the enterprise in question. Furthermore, he says, economists tend to have a sense of historical mission. Economists see their discipline as a recent and necessary development in intellectual history, away from the superstitions and the mysteries of the past. The science of economics is yet another step in our progressive confidence that we humans can measure things dispassionately and thereby come to understand how they really work. Not uncoincidentally, this confidence has grown during the last two hundred years, when a great culture transformation has been taking place. The percentage of the world's population that farms for a living has fallen from 82 percent to 3 percent (at least in the West). Most of those who have moved off the farm have moved into industry in one way or another. Furthermore, this transformation has taken place precisely because agriculture itself has become more industrial, using increasingly sophisticated tools and techniques. Thus, Dahlberg reasons, it seems natural to some people to apply newly evolved economic measurements to agriculture as if it had also evolved into something completely different from what it has always been—a flexible industry instead of an inescapable way of life. As a consequence, Dahlberg says, economists tend toward a "belief in man's dominion over nature" to share a "curious ambivalence toward peasant agriculture" along with a "systematic lack of concern with the environmental costs of agriculture" (p. 12).

Whatever one thinks of Dahlberg's bold analysis of what he considers the mistaken assumption of most ag-economists, it helps to explain why the proponents of a protectionist policy think the way they do. *They* think that agriculture has not really evolved into an industry, even as it has become more industrial, because they think the link binding the farmer's work to the life of the soil has not been broken. The industrialization of agriculture, although real, has been a change in degree, not in kind. Agriculture is still agri- or field-culture, with its needs rooted to the soil as well as being entangled in the marketplace. Only by misunderstanding this could an economist make the mistake of applying to it free-market standards. That is why the protectionists in turn argue for something like what Dahlberg calls a "contextual model" in measuring agriculture, by which he means a model that could comprehend the whole context in which agriculture takes place. That means recognizing that "agriculture mediates between natural and man-made systems and is strongly influenced by local cultural beliefs and views of man's relationship to nature" (p. 18). Using such a model is tantamount to advocating a "protectionist" policy for agriculture; the government must recognize agriculture's special and delicate nature as a mediator between the world of plants and animals and the world of human beings and must move to protect her from the rougher players in the free-market economy of industry.

Dahlberg puts equal emphasis on the culture and the nature of agriculture. Every farm is rooted in its particular culture, he says, whether it be a farm in the American Midwest or the Middle East. Patterns of land ownership, farm size, access to markets, and the distribution of the larger culture's wealth to its farming citizens are the way they are for more than economic reasons. The history of a nation's migration, the religion of its farmers, and the level of education in rural areas all play a part; but where a thinker such as Schultz might tend to think of these cultural factors as potential entanglements hindering large-scale economic change, Dahlberg sees them as part of an essential web binding farmers to their soil, their fellows, and their fellow countrymen. One cuts through such webs at the peril of the health of the whole. He lists many programs established by well-meaning foundations such as Ford and Kellogg with the purpose of modernizing peasant agricultures throughout the world, in Mexico in particular. In Dahlberg's opinion, such programs have often been insensitive to the social-economic web of local farming. They have encouraged the use of modern equipment and knowhow that has in turn favored the large-scale farm best suited to the tools and consequently brought about widespread social change that has driven many small farmers off their land. The resulting farms are often oriented more to filling city stomachs or foreign stomachs than local stomachs. In general these programs have been "landlord-biased" rather than "peasant-biased"; and although Dahlberg is speaking about peasant agriculture specifically, his remarks could extend to small-scale family farming in Western nations as well. He is talking about the tendency in Western governments to favor a policy of cheap food over a policy of good farms. The choice has clear consequences. The first is more economical in the strict sense of the term; but it entails the loss of local culture. One simply has to decide whether the loss is worth the gain, and this decision is, at root, emotional rather than economic.

Dahlberg also points to economic rationales for what is essentially a moral argument. Granted, the web binding the farm to its own soil is a delicate one, expensive but essential to maintain. Still, local, conservationist, and ecological methods are going to be more economical in the long run. For example, it may be cheaper in the long run to bring fertil-izer onto fields in the form of animal manure than in the form of chemi-cal fertilizer. Even if a pound of manure appears to cost more than an equivalent weight of chemical fertilizer (with the same N-P-K value), it might not cost more if all the production costs were factored in. With manure one must own and maintain an animal and often a spreader; with chemical fertilizer someone must own a patent often developed with government funds, have access to raw materials, and own the means of production and distribution. These "maintenance" costs can be high (p. 80). When all these factors are figured in, perhaps the cost of the two fertilizers is not that much different. As George Borgstrom points out in his book *The Food and People Dilemma,* no farmer run-ning the breeder-barn of a verticalized chicken business really feeds 75,000 chickens daily all by himself. Rather he or she is aided by those who make the equipment, raise the grain, process and package the feed, make and install the feed conveyor belts, and perform other related ser-vices. "The U.S. Department of Agriculture states that in 1947, 5 million people were engaged in supplying farmers. By 1965 this figure had risen to 7 million" (Borgstrom, p. 88; recent USDA estimates put the figure closer to 12 million). Beyond the question of infrastructure, the price of raw materials can change very quickly as was the case with the price of oil in the middle 1970s. Suddenly the price for artificial additives might soar just at a time when a country's farming had become fully hooked on chemical farming.

This is the argument that in the long run, decentralized, diversified farms are better for the health of the soil, which provides the bedrock for all future economic health. They resist disease better. Dahlberg crit-icizes what he calls the "highly simplified agriculture systems" of the present day precisely because they are susceptible to plagues and epi-demics in addition to social-economic disruptions. He admits that all ag-riculture simplifies the environment in order to produce a particular food in abundance. The difference is that industrial agriculture allows simplification to go too far. The same germ plasm with its limited resources of resistance appears in the hearts of millions of contiguous acres of a single crop. If the climate unexpectedly changes and damages the crop that a myriad of farmers have selected as the most productive in their wide area, then the whole crop might be lost. The basic idea here is, if nature hates a vacuum, then it also dislikes a monoculture. She does ev-erything she can to return a farmer's field to the complexity of nature in the raw. The best defense against this natural tendency is a kind of ju-jitsu, whereby the defender turns the attacker's own strength against the attacker. This means that agriculture must divert the hunger of bacteria and insects in many directions at once, not just to the table of the mono-culture feast modern scientific agriculture wants to supply. This means that we must enlist many natural predators in the struggle to protect our fields; the extinction of any plant, animal, bird, fish, or insect in the environment should sound an alarm bell that perhaps our agricultural system is at peril. We simply can never know what natural allies we shall need in the future, so we cannot afford to lose a single one in the present. *"Let us keep our evolutionary options open,"* Dahlberg says with emphasis (p. 153).

Dahlberg advocates a contextual model, as he calls it. We need to measure "the global availability of air, water, and soil (past, present and future)" before we make any policy decisions about agriculture. We need to estimate our reserves, estimate what must remain in reserve, and monitor their use in detail; then we need to develop a policy that will hold the reserves intact, which Dahlberg believes will be a good farm policy before it will be a cheap food policy. E. F. Schumacher provides guidelines in his comments about the need to develop "intermediate technology"—"vastly superior to the primitive technology of bygone ages but at the same time much simpler, cheaper, and freer than the super-technology of the rich" (quoted in Dahlberg, p. 166); this would enlist a great diversity of farmers in the diverting of nature's jealous hunger from our fields.

We can find many interesting statistics in George Borgstrom's book to back up Dahlberg's arguments about the hidden costs of industrialized agriculture: "through tractors the average American burns 4.5 million BTU's or 32 gallons of gasoline per capita" (p. 89); factoring in all energy inputs, "the feeding of each American requires the equivalent of 150 gallons of fuel per year" (p. 90). Another way to put it is that it costs 12,000 calories of energy per day to provide each American with the 3,300 calories he or she on average consumes. This is a huge but hidden deficit economy. In fact, even the amount of energy used in processing food exceeds the amount of caloric energy it supplies, not even taking into account the amount of energy it takes to grow the food in the first place (p. 91). These statistics alarm Borgstrom greatly: "While one might argue that this is a reasonable price to pay for the invaluable fringe benefits of special fats and protein, it is far more essential to realize that this practice places the future of this kind of agricultural operation in jeopardy" (p. 93). He believes with Dahlberg that the long-term prospects are not good, and in fact they are not economically sensible, to say nothing about not being ethical: "The excessive energy use by the well-to-do world may therefore be nothing more than a grand-scale swindle, creating an empty purse for the Hungry World" (p. 94).

Like Dahlberg and Schumacher, Borgstrom believes that the choice is not between old-fashioned farming and newfangled technology, but a choice that lies somewhere in between—an appropriate kind of farming that factors both short- and long-term economics in with its careful assessments of continuing soil health. This requires education of the masses who live in the cities and who need to appreciate what the true costs of cheap food might be; they need to be informed about the value (and the cost) of good, nutritious food, especially in the age of fast food; their political support is needed to fund a more diversified farming that will provide wholesome food, while working with and not against nature's own insatiable hungers. We need to monitor closely the chemicals that might be sneaking into our food from artificial pesticides and fertilizers. We need to monitor the chemicals that are building up in the soil and in our dumps as the result of our residues the natural decaying process cannot absorb. He cites a compelling example:

> Many cities in central Europe and the southern part of the Soviet Union recycle their sewage by cultivating fish in ponds, frequently in combination with the raising of ducks. Several Southeast Asian cities do the same and obtain fine gouramis, carp and other fish which all feed on worms, plankton and living organisms nourished by the riches of sewage. If we are not to be submerged by excessive expenses, the recycling of organic matter will have to be our future pattern. It is an urgent matter that food be shunted into such production chains (p. 102).

Borgstrom finds himself particularly alarmed by the massive wastes produced by what he calls the "broiler hotels, the hog factories, and above all the feedlots" of verticalized agribusiness. These operators produce in a small space more waste than can be done away with efficiently, and greatly pollute their environment. This needs to be recycled, he says.

Basically, Borgstrom says we need an agricultural policy as sensitive as it is broad. We need to harvest protein more efficiently and harvest waste more effectively, and the cost of both must be factored precisely into the cost of food in the supermarket.

Vernon Gill Carter and Tom Dale come to similar conclusions after taking an even broader view in their book *Topsoil and Civilization*. They consider the history of soil use from the earliest traces of human farming to the current practice in the Western world, and they are not happy with what they find. One civilization after another has risen to power by mining its resources and collapsed when those resources ran out. Agricultural resources have been the most critical to this rise and fall. The process basically works by a culture's progress depending, at first, on progress in agriculture. Agriculture must advance to a certain level of sophistication in order to free the large numbers of workers and craftspeople and politicians a culture needs to become a great world power. But if agriculture goes too far, to the point where it mines the soil in the process of feeding the multitudes, then the collapse of the barn pulls down the temple and the capital with it:

> With the advent of civilized man, about six thousand years ago, the soil building process was reversed in most areas where he resided: the quantity and quality of soil and the amount of life the soil supported all began to decline. His superior tools and intelligence enabled civilized man to domesticate or destroy a great part of the plant and animal life around him. But more impor-

tant, his improved tools and techniques helped him, unwittingly, to destroy the productivity of the soil that supported life. His intelligence and versatility made it possible for him to do something no other animals had ever been able to do—greatly alter his environment and still survive and multiply (p. 6).

The survival of mankind has not been the same thing as the survival of nations, of course. When the soil of the great nations of the past ran out, its farmers became colonists of other lands, leaving behind them eroded soils. In most of Greece, for example, and some places in ancient Rome and Egypt, the soil has not recovered to this day. Greece is now known for its spectacular, rocky, barren landscape; it used to be green and lush until goats and sheep gnawed its foliage beyond recovery. Most of the poor people in the third world are poor because their ancestors depleted the rich topsoil long ago. The question is, What do we do now that there is not other empty land to migrate to and cultivate when the fertility and the topsoil of a country's land disappears? There is no place to go but back to the same field year after year; and the authors are not optimistic that any serious effort is being made to make agriculture truly sustainable. Our agricultural schools are not teaching the "homely fundamentals" about water supply, tilth, conservation, and beneficial wildlife—that these resources are not static but are changing continuously, and only an acute farmer can adjust his farming to their changes in order to produce a crop year after year in the same place. Only such a crop can be counted on to keep the distant city's food bins full.

Carter and Dale are hopeful about America's prospects when they consider the great wisdom every culture has had, including our own, about how to conserve and enrich the native land. They despair when they consider what seems to them the lack of political will to pursue a good farm policy in this country at this time. The time is critical:

> Most of our natural resources have already been depleted to an alarming extent. Modern technology and enterprise have made it possible for us both to develop and to exploit these resources faster than any other people in history. The United States is now approaching the state in resource depletion at which many of the past empires and civilizations started to decline" (p. 24).

What is needed is the political will to pursue the policies needed so that the waste and the damage be stopped. Fine the polluters and the eroders; fund the conservation services; alert the public to the high cost but high worth of long-term conservation. Only a good farm policy will protect and preserve our soils.

Conclusion

The debate arranged in this review urges us to take a stand: should we choose a cheap food or a good farm policy? Each side claims we need to move further and faster in pursuit of one policy or the other. But does their continuing disagreement about where we should be going imply that right now we are someplace in the middle? When taken together, all the uncomfortable half-measures that make every participant in the policy debate anxious—the protectionist subsidies coexisting with free-market permissiveness—do they comprise a working compromise between them? Do we have right now a rough, *de facto* "good food–good farm" policy? Evidence that we do is that none of the contenders in the arena of agricultural policy ever seem satisfied. Is this a good sign?

Readings

Borgstrom, George. "The Energy Swindle"; "Food in Man's Ecology." Chaps. 8 and 9 in *The Food and People Dilemma.* Belmont, Cal.: Duxbury Press, 1973.

Carter, Vernon Gill, and Tom Dale. "An Overview"; "Can the U.S.A. Survive?" Chaps. 1 and 13 in *Topsoil and Civilization.* Norman: University of Oklahoma Press, 1974.

Dahlberg, Kenneth. "On the Ecology of Theories"; "The 'New' Seeds and the Logic of their Growth (Or Jack and the Beanstalk Revisited"; "New Approaches to the Future." Chaps. 1, 3, and 5 in *Beyond the Green Revolution.* New York: Plenum Press, 1979.

Easterbrook, Gregg. "Making Sense of Agriculture." *The Atlantic* (July 1985), pp. 63–78.

Heady, Earl O., et al. "A Picture of Agriculture in 1980." Chap. 12 in *Roots of the Farm Problem.* Ames: Iowa State University Press, 1965.

James, Gwyn. "Reflections on Agricultural Policy." Chap. 9 in *Agricultural Policy in Wealthy Countries.* Sydney: Angus and Robertson, 1971.

Peterson, E. Wesley F. "Agricultural Structure and Economic Adjustment." *Agriculture and Human Values* 3, no. 4 (Fall 1986): 6–16.

Chapter 4

Agricultural Techniques

The Cases

As American as Apple Pie

For six generations the Vorst family has run an apple orchard in upstate New York. Traditionally the family has been large, with some generations having as many as twelve children growing up on the orchard and working at their many chores from a very early age. Many of the children over the years have moved away from the homestead orchard and established other orchards in the area using the family name. A "Vorst" apple has become known throughout the state as a quality eating and baking apple.

Harvest time has always been the busiest time at a Vorst orchard, as it is at any orchard, and the family has always hired itinerant workers to help them with the picking and boxing of apples. Workers have always enjoyed working at the Vorst orchards because the Vorsts have extended their sense of family to the people who work for them seasonally. They have provided good, clean temporary housing with running water. Because their other labor costs are low, with the large number of family members who work without salary, the Vorsts have paid their workers higher wages than other orchards and have still been able to market cheaper, quality apples. As a consequence, the seasonal workers the Vorsts have hired over the years have tended to be the same families, loyal to their employers and dedicated to doing good work for good pay.

The current generation of Vorsts running the homestead orchard is different from previous generations, however. They are worried about the future of their apple business. For the first time they are facing stiff competition even running their farm fruit stand. For three years running, apples grown in Oregon and made available locally for half the price the Vorsts need to sell their apples to show a profit have been slowly closing the doors of local and state markets for their apples. During the first and second year of the competition, the Vorsts thought that the Oregon apples were cheaper only because of an unusually good growing year on the West Coast coinciding with an unusually bad growing year on the East Coast. But now after studying the situation closely, they have decided that the competition is going to be permanent. The Oregon Apple Growers Association has shrewdly developed transportation routes and advertising campaigns to help market a product that is being produced more cheaply than the Vorsts can hope to. The main reason why their production costs are lower is the association's development, working closely with an agricultural chemical company, of a new chemical spray that forces all the apples on a tree to ripen at the same time, thus making single-pass picking possible. The overall strategy of the association seems to be to take control of the eastern market from eastern growers. The Vorsts suspect that perhaps the Oregon State Agricultural Board is subsidizing production costs as a part of the strategy.

The Vorsts meet together with their family members on adjoining orchards and discuss alternative business strategies. They whittle their choices down to two. Each orchard can try scaling down their operations by cutting their hired labor pool, harvesting with mainly family picking teams that would work together, and selling only in nearby regional markets so as to cut transportation costs. This way they calculate they could market a cheaper apple than an Oregon apple; and this way they could maintain the spirit of their family-run business while still making a profit. The other alternative would be to purchase the new apple-picking machines just coming on the market. They were developed at Cornell University, the state's land-grant university, supported by funds from several major tractor-making companies. These machines ride over the trees and gently knock apples off their branches with dangling rods. The machines must be used along with chemical sprays that cause all the apples on a tree to ripen at the same time. These machines are expensive. But if all the Vorst family orchards incorporated into a single large farm, they could afford to sell to real estate developers some of the land that is now used for seasonal labor housing on each farm. They could use this money to make down payments on the machines. They calculate that within two years these machines would pay for themselves by lowering labor costs, and that after that, they would essentially pick the apples for free. The Vorsts could begin marketing cheaper apples immediately, even at a loss, before the machines began to make the lower prices profitable, if each orchard were willing to contribute to a fund that would subsidize the orchards that would suffer the most in the group. The Vorsts decide to incorporate under the name Vopco, Inc., and buy the machines. They reason that even if they decided to scale back operations and continue as a "Mom and Pop" apple farm, aggressive competition from out of state might eventually force them out of even regional markets. They are going to have to fight fire with fire.

The people most disappointed by the decision to form Vopco are the farm workers who have worked with family members for years and who know now they will never be hired to pick apples again. Particularly disappointed is one family with several children in college whose tuition was being paid for by the wages of many older family members. Few of them could speak English well, and none of them had ever attended school in the United States.

Corn Ball

The Corn Growers Association in a large midwestern state is proud of pointing out that its state produces more corn a year than every other country in the world besides the United States, Canada, China, and Russia. The state sells more corn to Russia than to any other country in the world. The state is good for growing corn. The land is naturally flat, the soils are deep from having once been a short-grass prairie, and the climate allows for a long growing season with lots of rain. Yields of sixty bushels an acre were common even before the age of modern chemicals and machines; nowadays some particularly well-situated and well-run farms can produce three hundred bushels an acre.

But the Corn Growers Association officers believe that their members should not become complacent. Too many of them use what is now considered traditional amounts of chemicals on their corn and drive tractors and combines that are as much as a decade out of date, given the continuing technical sophistication of agricultural engineering. The officers decide to stage a series of regional contests to encourage bushels-per-acre yield increases. Participating farmers will set aside a ten-acre test plot and carefully record the combination of chemicals, the array of techniques used, and the dates of application on that plot during a growing season. This information will be coordinated with weather data during the growing season. At harvest time each farmer will keep secret his or her yield on the test plot to build suspense for the annual meeting when the farm with the highest yield will be announced. At the same time, each participating farmer will receive a record book showing the yields on each farm, dates of application, and the configuration of chemicals and machines that were used. The officers of the association figure that this way they will sharpen the natural competitiveness among farmers who will see what other farmers are doing and plan their own future strategies accordingly. Farmers will thus be encouraged to learn from each other and to make the purchases of chemicals and machinery that will perhaps enable them to win the handsome seven-foot golden cornstalk trophy for themselves next time.

The contest is successful, and for a number of years yields per acre increase statewide at a substantial rate. This helps some farmers who are hurting from a world-wide surplus of corn that has driven the price for a bushel of corn in some places well below the cost of producing one.

But some of the farmers in the association begin to grumble that there has to be some new way of bringing production costs down. The prices for the chemicals and machines keep going up as they become more sophisticated, while the price of corn seems to keep going down as more and more chemicals and machines produce ever greater yields. Aware of this grumbling, a new company, Blanton Agricultural Aviation, known as Baa, Inc., sends a sales representative to address the annual meeting of the association. The representative explains that his company has a new service. For a fee, Baa will fly a plane over a farmer's fields before, during, and after the growing seasons, taking special infra-red photographs that help the company's researchers analyze just where the farmer needs to put precise applications of chemicals, irrigation water, and precisely where to cultivate just the right amount. This will enable the farmers to fine-tune their use of chemicals and machinery and thus make them more efficient and lower production costs. The new service will make the annual contest even more competitive.

Association members discuss the proposal. Some, including the association's officers, want to take advantage of the new service, because they see it as a logical extension of their long-term committment to progressive, modern, and therefore technical farming. But a small group of younger farmers is adamantly opposed. They argue that Baa will force them to become even more dependent on technology for the growing of their corn. They will lose even more control of the costs of production—especially if the new air service becomes absolutely necessary for each farm to remain competitive, not just in the contest, but in the overall yields of the farm. They argue that the need for the new air service points out the failure of the whole new chemical system of farming, in that the system becomes ever more complicated to keep working at all. At the same time, it produces the enormous yields that keep the price of corn low.

The younger farmers argue that corn farming instead ought to return to the system of crop and animal rotation something like their grandfathers employed. Different grains were rotated on a field in different years, with grain crop rotation coordinated with hay and animal pasture rotation. This way animal manures and green manures would provide fertilizer along with improved tilth and soil texture. Changing crops would frustrate the insect populations that fed on each. Such a system would require putting up fences again and decreasing field size, and it would complicate grain farming some. But it would not have to be a throwback to the old days. The tried-and-true methods of rotation could be brought up-to-date with new research, new seed hybrids, new breeds of cattle, and selected use of chemicals and new machines to enhance the effects of natural fertilization and integrated pest control. Perhaps farm sizes would get smaller as more farmers were required to run the new system. Each farmer would have to reduce the size of his or her corn production in order to be able to keep on top of the whole more complicated enterprise. Each farmer might take better care of the farm this way; and because production costs might come down, and because yields would certainly come down, the price might go up. It was possible that a nice balance could be reached this way between production cost and price that would ensure an honest profit for honest work. At any rate, such a system would bring control back to the farmers deciding how to farm and take it away from the researchers and manufacturers and even airplane pilots who now seemed to determine how corn was to be grown. It would be critical, of course, that many farmers made the change over to the new system at the same time so as to equalize the competition and to give all the participating farmers time to bring their production costs into line with the higher prices for lower yields.

The older association members listen politely with some amusement, especially to the arguments in favor of crop rotations. "Sounds like a lot of work, rotating those animals around my fields," one farmer named Vlad Maxwell joked during refreshments, "don't know as I can get back in time from Florida in the spring to do it." Another farmer remarked more seriously, "Sounds like socialism to me, what with all of us going back to the old ways together." At the end of the meeting, a majority of association members voted to add the services of Baa, Inc., to their contest game plan and to enlist airpower in the struggle for agricultural efficiency.

A Pole Apart

The year is 1825 and a young Polish farmer Wojcik has just emigrated to the New World in hopes of a better way of life. He settles in what was then the Wisconsin Territory probably because, one of his descendants was later to joke, "he kept going west until he found soil as poor as the soil he had left behind—it was all he knew how to farm." Actually it is more likely he settled where he did because there was a small community of Polish farmers that had already settled in the area, and although the soil was poor, it was cheap, and therefore all that the new immigrants could afford. The region in which he settled had the advantage of lying adjacent to the new settlement called Milwaukee, where there were markets for the grain and beef that the new settlers could grow. Wojcik begins his life as a farmer by clearing ten acres of land that he divides into a vegetable plot for the house, a small grain field, and a pasture for his animals.

Although Wojcik is new to America, he is not new to farming, having grown up on a farm in Poland that had been in the family for many generations. He had learned from his father many tricks of the trade, such as the simple one of interchanging the fields in which one kept animals one year with the fields in which one grew grain. This way the manure the animals left on the field one year could be used as fertilizer for the grain crops the next. Wojcik uses this method during the first few years he farms. His fellow farmers notice what he is doing and tell him that in America no one has to farm that way anymore. The topsoil is thick, even here on the sandy soils near Lake Michigan, and has its own fertilizer packed into every handful. Any farmer can afford to own enormous amounts of land in the new country, and it is less work and therefore cheaper to keep clearing land for new crops as old land wears out, than it is to put up and maintain fences and to plow up pasture land every year for the rotation. Wojcik listens to what his fellow farmers say and decides he does not want to abandon the old country ways of farming that he has brought with him. His fellow farmers start telling jokes about his old country ways.

Wojcik and his family keep to themselves for the most part, not socializing with the other Polish farm families in the area very much. They travel to the city whenever they can and soon begin to work very hard on improving their English. Wojcik hears about a new Agricultural Improvement Society that has formed in the city, mainly made up of doctors and lawyers and other educated people who are interested in improving local farming around Milwaukee. They hope that the price of locally grown food can remain low, its supply constant, and its quality high. The society has been corresponding with similar societies in England that have been disseminating information about what is called the new English Farming Methods. These methods are dedicated to improving English agriculture in the years after American independence. The former mother country found it could no longer depend on cheap food imports from its colonies to keep the price of food low when there was no longer a place for English farmers to migrate when their own lands ceased to be productive. Wojcik learns that the new English farming methods are somewhat different from the methods he learned from his father in Poland. The farmer still uses rotations for natural fertilization, pest control, and soil building. But the rotations are complicated, requiring sequential plantings of different grains in different years, fallow years, and pasturing years for different soils. For example, an expert at the local society helps Wojcik to decide that he needs a rotation of oats-oats-corn-wheat-hay-hay and animal pasture in a seven-year cycle for each separate field he works. He can have several different rotations at different stages on different fields, so that he can produce the same amount of crops each year on the farm overall. He must in addition spread manure from his barns on the fields that he is preparing for grain each spring.

After several years his fellow farmers notice what Wojcik is doing and find him very entertaining at first. He is working harder than they are keeping each of his fields going, and yet is making less money than they are with the smaller yields of the beef and grain crops that bring the most money in Milwaukee. But after some time, many of the farmers become annoyed when his new farming starts to interest some of their young sons who think there might be some point to farming the same fields over and over again, not only until their original astonishing fertility runs out. Some of the younger farmers hang around Wojcik's farm asking questions and find themselves traveling to Milwaukee to attend the lectures of the Agricultural Improvement Society. Eventually the local farming community becomes polarized. The older farmers farming the new American way have very little to do socially with the younger farmers farming the new English way. The split worsens as a trend begins whereby younger progressive farmers begin buying up older farms to farm them the new way when the older farmers find they must move on to new territories further west to find new fertile ground.

Wisconsin joins the Union in 1848. When Congress passes the Morrill Act some twenty years later, establishing a land-grant university system to study methods of progressive farming, members of the Milwaukee Agricultural Improvement Society form the nucleus of a new agricultural experiment faculty at the University of Wisconsin at Madison. The first new agricultural research building at the university is named after Farmer Wojcik, by now a legendary pioneer of progressive farming.

Fly Away

The Vitek Dairy Farm in upstate New York has a wide reputation for producing wholesome, rich milk. The Viteks sell milk to a local dairy co-op that markets the milk of a dozen farmers under a single co-op label. They also sell milk under their own label from a small shop set up by the side of the road that leads to their farm. Many neighbors and people from a nearby town make special trips to the shop to buy the Viteks' own brand, perhaps because the Viteks mix cream-rich Jersey milk with thinner Holstein milk and produce an especially flavorsome but still relatively fat-free milk. Some good friends and nearby neighbors bring by their own gallon glass jugs to be filled with raw, unpasteurized milk that the Viteks sell on the side. The general feeling is that Vitek family members drink their own raw milk, and they appear as healthy as bulls. What is good for the farmer should be good for the consumer.

One spring unusually heavy rains saturate the pastures where the Vitek herd will spend the summer, filling many hollows with puddles and swelling small brooks and ponds to overflowing. Insect populations thrive and bedevil the herd with sucking flies swarming about their eyes and skin wounds and biting flies such as horn flies and horseflies all over their hides. Milk production drops in the herd that becomes nervous and edgy with many small infections on their skin. Life also becomes miserable for the Viteks themselves who take to dousing themselves with insect repellent kept on a shelf right beside the door.

In past summers, even after wet springs, the early fly populations usually died down to acceptable levels. Insect-eating birds return, the land dries, and steady winds brush over the animals and hilly pastures, waving the surviving insects away. The Viteks pride themselves on their unwillingness to use pesticides of any kind on their farm. Partly they are concerned about the effect chemicals that get into the animals' systems might have on the milk or what residues might do after leaching into the watershed. This year, however, the rains continue, and the air remains sullen and close. The Viteks themselves become nervous and edgy. They have great affection for their animals, and their discomfort at seeing them suffer finally overcomes their reluctance to using chemical pesticides. They contact their local extension agent for advice.

The agent suggests they use a chemical spray known as Vapona, whose active ingredient is dichlorvos, the same chemical used in popular brand-name repellents. In high concentrations the chemical is highly toxic. But if applied carefully, well away from buildings, with gentle spraying motions over the animals' bodies, the chemical is safe and highly effective in repelling insects for several weeks after each application.

One of the Viteks is a young man in his twenties who has become a vegetarian and will not eat even the beef the Vitek farm slaughters itself from its own culled animals. As a teenager he had convinced his parents to close down their veal calf operation because he found it repulsive to make healthy young animals anemic for an early slaughter. The Viteks now sell their extra bull calves to a neighbor who still produces veal. The Vitek son no longer lives or works on the farm, having "flown the coop," as his mother puts it, because he could no longer feel comfortable even with the idea of keeping contented cows behind fences. He teaches philosophy at a nearby college where a colleague in the biology department, hearing from him of the problem on the Vitek farm, tells him about IPM or integrated pest management. The biologist puts him in touch with a small local company that sells parasitic wasps to prey on flies. A salesman for Merry Insects explains the process. An adult wasp (*Spalangia endius*) deposits eggs in the larvae of flies. The eggs hatch and the young wasps eat the larvae as their first meal. Adult wasps emerge and continue the same process, so that an initial release of wasps by the company begins a biological chain reaction, producing more and more wasps until the prey species is effectively controlled.

Young Vitek is impressed and calls his parents with the idea when they are considering the extension agent's advice. They are intrigued and call the agent to ask him to compare the two strategies. The agent says he is aware of IPM, but he does not recommend it to farmers because it is never as effective as chemical controls. For one thing, chemicals work immediately, whereas it takes several weeks before predator-wasp populations become high enough to begin having a noticeable effect on fly populations. In the meantime the cattle continue to suffer. Then, after an initial kill-back, predator and prey populations achieve a balance, he explains. Predators can never completely eradicate their prey without eating themselves out of a livelihood. Thus, after a few months, there will always be flies in the air with IPM. Chemical controls, on the other hand, repel all the pests all at once.

The older Viteks find the extension agent convincing. But when they talk to their younger son, they find him adamantly opposed to the use of chemical sprays. Out of an affectionate respect for his opinion, they decide to hire Merry Insects to control their flies. They figure that if IPM does not work to their satisfaction, they can always try chemical insecticides later.

Keep Away

A year later the Viteks read an article in a New York State Extension Bulletin about a new feed additive developed at Cornell University, New York's land-grant university. It is called bGH, or Bovine Growth Hormone. The hormone occurs naturally in all cattle. It is released by glands as cows mature to regulate their lactation. Scientists at Cornell have been able to synthesize large amounts of the hormone which, when fed to already mature cows, stimulates their milk production significantly. Scientists estimate that it will increase the productivity of individual cows between 15 and 30 percent, with no diminishment in quality of the milk and no adverse side effects for the animal.

They find this news alarming. Mr. Vitek estimates that with more milk being produced the price of milk will fall even lower than it is now. Currently they make a profit at their business only because the government pays them a subsidy of twenty cents a hundredweight over the wholesale price they get for their milk. The government also buys up vast stocks of surplus butter and cheese that dairies produce for distribution to the poor or for more or less permanent cold storage. He writes his congressman, Donald Seybold, urging him to block the commercial introduction of bGH, arguing that it will simply increase the amount of subsidies the government will have to pay to make up the difference between what milk is worth and what a farmer needs to live.

Congressman Seybold responds in a letter in which he says he sympathizes with Mr. Vitek's position. But he says his office predicts that the introduction of bGH will help lower subsidies. More productive cows will mean that farmers will be able to produce as much as they do now with fewer animals and thus be able to lower their production costs of housing, feed, and pasture. Bringing down the price of producing milk will bring up the farmer's profit from its sale. The congressman tells Mr. Vitek he remains a staunch supporter of agricultural research in general and bGH in particular.

The Issues

Farm tools do not speak when they are hanging from a shed wall like a shovel or a hoe, or parked in an old wooden barn like a tractor, or piled up in a new aluminum-sided storage building like a stack of chemical fertilizer bags—or even when whirring quietly in the farmer's farm office like a new computer she is using to keep track of pork belly futures on the Commodity Exchange. All tools are dumb and apparently innocent of any of the uses to which they are put. The tractor will plow in manure with the same dispatch that it will plow in chemical fertilizer. This truism might lead one to believe that one should go talk to the farmer, not the tractor, if one has a moral conviction that one kind of fertilizer would be better for the soil, for the farm, or even for America. It seems it is only in the farmer's mind where the decision will be made how to use a tool, or whether to use one tool or another. Therefore it is not surprising to find in this chapter that the discussion about the technology of farming—what its tools should do—is going to repeat many of the arguments used in the discussions about principles and about policy that we have already reviewed. Now the discussion will be brought down to earth—the particular earth of the particular farm where a farmer has to decide here and now how to farm the farm. Here is where principles and policy dig in.

But there is something new that distinguishes the arguments made about the use of technology in farming. It is the conviction among progressives (as they were defined in chapter 1) that technological progress cannot be stopped, no matter what individual farmers think about tools. Technology has its own mind, as it were, and it is bent upon its own continual improvement. The romanticist E. F. Schumacher, although he inspires Kenneth Dahlberg with his talk about smaller machines for smaller farms, is really whistling in the wind. Farmers *must* use better, bigger, and more sophisticated machines to keep their farming efficient.

To this way of thinking, it is the integrated system of technological development that makes it inexorable. First, individual discoveries occur in scientific and engineering labs. Then technicians tinker with designs and procedures on the shop floor. Experimenters test new techniques in the field. Agrieducators, extension agents, and salespeople inform progressive farmers of promising new techniques. The farmers apply the acid test of commercial profitability and offer suggestions of their own back up the line. To complete the circle, farmers made prosperous by the new techniques and the agribusiness concerns made profitable fund the basic research done in academic and company laboratories. They provide the seed stock. Thus, the system keeps progressing because it serves the self-interests of all its participants. Furthermore, the larger society offers financial support and other encouragements because of the benefits it gains from lower food prices, greater varieties of food, and a smaller, more efficient farming community. From this perspective, discussions about principles and policy are really no more than rationalizations, justifications, or criticisms of what technology has already wrought.

Sustainers respond by saying that technological development is not inevitable. It proceeds by the deliberate direction given to it. Too often that direction is supplied by individuals greedy for personal profits regardless of harm done to the commonweal. Instead, society ought to direct the development of its technology in order to serve social and ecological needs in addition to purely economic needs.

Sustainers are vulnerable in this debate on several counts. They tend to share the assumptions of their opponents that technological development, left to itself in a capitalistic society, will tend toward the development of large-scale techniques. These techniques concentrate the ownership of the means of production—industrial as well as agricultural—into fewer and fewer hands. Furthermore, it is inevitable that public universities will play a large role in developing the technology of scale and the economics of concentration. The system described above will work its will. Thus there does seem to be an inherent, progressive momentum to technological progress.

But they make the counterargument that it is therefore all the more urgent to redirect this more or less natural tendency in order that it will do the greatest good for the greatest number of citizens, farmers, and consumers alike. Sometimes this will mean developing the science and the technology of smaller scale, more diversified farming tailored to the culture of the farmers and the landscape of their farms. Perhaps this will mean that the government will fund research projects intended to help the smaller farm if it turns out that corporate research favors the farm run more like a corporation than a garden. Perhaps this will mean a good food policy that might reward the manufacturing of small manure spreaders by promising a market among the farmers who are working to supply the market among consumers for local, organically fertilized products. But in any case it means that decisions about how to use technology need to be considered apart from technology itself at every stage of planning, research, and application.

To go along with their faith in supplying a socially conscious direction to technological development, sustainers are more likely than progressives to assume that there is no such thing as an objective science or a neutral technology—dumb, innocent tools—on the farm or anyplace else. Science and technology are always practiced for a purpose. This is not to say that a scientist or a technologist always knows ahead of time what the practical use of a discovery or invention might be. But it is clear that, first of all, science and technology develop in the interests of change. They focus on a different future. Furthermore, most scientists and technologists realize what kind of change they are working for. It takes no special vision for anyone to recognize that a $200,000 fruit picker favors a certain kind of farming, in the same way that a cheap hand gun favors certain activities. It is not unprogressive to think a society might benefit from regulating the manufacture and use of each in line with its other goals and social concerns. Recall Theodore Schultz's idea that technology is never an isolated development or a force in agriculture. New techniques change everything. The land, crops, even the farmer has to change when new technologies come into use. Conversely, a sustainer might argue, new perceptions about social and environmental needs should dictate what the new techniques should be.

On both sides of the issue, the writers reviewed here draw on the principles and the policies discussed in the previous two chapters. In addition they draw on broader discussions of the appropriate use of technology being raised throughout our culture today. There is much that is familiar in the discussion, even if the arrangement is new.

But finally there is something special in the discussion of agricultural technology that we find nowhere else in public discourse today. It is the passionate belief that somehow farming is different from any other human activity that uses tools. Farming is essentially moral. In this chapter we begin with three different views of the morality of agricultural technology. Clarence H. Danhof argues that nineteenth-century progressive American farmers were morally superior to the traditional farmers of their day. Their use of new techniques and tools allowed them to take better care of the land than traditional colonial farmers did. Milton Snodgrass and L. T. Wallace counter that the traditional farmers of the colonial period and early Republic were moral in their own way. They insisted on the freedom to live and farm as they wished. But like Danhof they believed that this kind of stubborn independence was outmoded in the modern world. Today a farmer has to adopt the modern technology of his neighbors to survive. In agriculture at least, progressive morality is the only morality that makes any economic sense. Jim Hightower disagrees with all three men. He argues that modern progressive and highly technological agricultural has lost its moral bearings and that we need to restore the split between an old-fashioned moral concern for the independent farmer and the new technology that farmer must use to compete in the marketplace.

The Review

Early Progressive Farming

We begin with the thoughtful, scholarly study by Clarence H. Danhof of the history of American agriculture during its most critical fifty years, from 1820 to 1870. Danhof traces the various progressive movements that led to their being institutionalized into the activities of land-grant universities by the Morrill Act of 1862. In his opinion, the Morrill Act is the natural flowering of a good plant with sturdy roots.

These roots were the years in which America came of age as a nation. She recovered from the two wars with England, spread her territory between two oceans, fixed her northern and southern borders, began to industrialize her cities, and encouraged farmers from Europe and her own eastern seaboard to begin settling the interior plains and Far West beyond the Rocky Mountains. At the end of the period, America fought and concluded a great civil war that established her social, political, and industrial future, as well as the future of her farming. All human enterprise from this time forward is tied up in the development of technology—especially agricultural technology.

Probably the most important concern of farmers during this period was the technology of land use. Until the beginning of the 1800s, land use entailed clearing a field with ax and fire, plowing it by hand or with horses and oxen, and then, after the field's fertility or topsoil thinned out, clearing new land. When new land became scarce in one locale, farmers moved to another where the soil still held its treasure of natural fertility from the long ages of trees and grass that had flourished there. Between 1820 and 1870, farmers began drifting west into what is now called the Midwest, breaking up the thick prairie sod to get at the incredible riches of fertile topsoil below. In some places its depth reached over thirty feet. The earliest yields of newly broken sod ground were astonishing to most of the farmers who immigrated to the new lands; and their early beliefs that its fertility would be inexhaustible seemed well founded.

But even the yields of these fields began to slacken after as little as five years of cultivation in some locations. By the 1840s declines were serious in some parts of Ohio, by the 1860s in Iowa. In reaction, some writers and other people close to American farming began thinking seriously about new techniques that would renew the fertility in a farm's field year after year. There were techniques available, with descriptions written in a language they could understand, that had been developed in England in the previous fifty years. England had lost her American colonies and their imports of food at the same time she was warring with her continental European neighbors and so had to turn herself to the intensive cultivation of her own home fields as her hungry population grew. Basically these were the techniques of applying animal or green manures to fields and rotating different crops in different years in the same fields, pasturing animals in the fields in some years and allowing the fields to remain fallow in others. Lime and marl were added to improve tilth and regulate the acidity of the soil. The English had experimented with different patterns of land use, and at the same time they had been developing new tools and animal-drawn machines to make the intensive use of fields and pastures more efficient. Increasingly American writers and thinkers urged that these patterns and techniques be used on American fields.

Some farmers listened to these urgings and began our tradition of "progressive farming" that Danhof believes continues to this day. But others did not listen, and according to him, their apathy and sometimes their protests continue to this day as well. Although Danhof takes a negative view of unprogressive farmers throughout history where others might distinguish a mindless resistance from a thoughtful critique of progressive agriculture, it is good for all concerned to be reminded that there is a long history of stubbornness and nostalgia in many efforts to resist the trend toward progressive farming.

The primary cause of the early resistance was habit. Many American farmers in the first decades of the 1800s and afterward were used to clearing land and moving on. They did not have a tradition of intensive agriculture for practical reasons. The initial effort of clearing land was very large and left little time and energy for intensive cultivation. Farmers were caught up in a vicious circle of never-ending labor: "While a number of years and perhaps a generation were required to clear the land held within a farm in forested areas, declines in productivity on the shallow upland soils of the East were sometimes so rapid that the rate at which new land could be brought into cultivation could barely sustain the total product of the farm" (p. 252). In addition, and as a consequence, animals were scarce on early American farms. There was little manure to spread on the fields. To these practical reasons could be added the often understandable tendencies of farmers to blame bad years and yields on the vagaries of the weather and insect infestations. It was hard for them to see the pattern of long-term fertility declines within the maze of many short-term patterns of luck and seasons. When some especially foresighted farmer would read an agricultural journal and try out some of the new techniques on his or her land, the results were frequently disappointing. For one reason or another, even a progressively inclined farmer would be convinced that farming had to be harsh to be farming at all: "Though there were important exceptions, most farmers, once established in a routine of cultivation, continued it unchanged throughout their careers" (p. 253). So strong was this force of habit that many farmers prided themselves on the ingenuity of their efforts to throw animal manure away, in some cases by building barns over creeks that would carry the matter downstream.

The countertrend toward progressive farming can be measured in part, Danhof says, by the increasing value placed on animal manures by farmers in southeastern Pennsylvania, southern New York, and eastern Massachusetts in the 1820s and in the older sections of New England and the Middle Atlantic states by the 1850s. Cellars began to be built under new animal barns to hold and preserve the value of their manure. This practice began to spread into Ohio and Indiana by the middle of the 1850s. Acceptance was slow: "A survey by the Indiana State Board of Agriculture in 1857 on the utilization of manure aroused little response, but those who did answer indicated that although waste was still typical, a change was taking place and manures were increasingly applied to the fields" (p. 260). The practice was widely established by 1870.

As the use of animal manures spread westward, so did the use of gypsum and lime, with imported guano used for fertilizer spreading from east to west after its successful introduction into the fields of Connecticut and Maryland beginning in 1848. Guano was followed by menhaden oil, and even chemical fertilizers in the form of superphosphate of lime in the early 1850s. The use of fertilizer often went hand in glove with the practice of crop and field rotation. Again the practice spread from east to west; again Indiana farmers resisted longer than farmers in other western states: "[They were] a decade or so behind those of Ohio in turning to crop rotations" (p. 274). But everywhere, along with these techniques, went the technology of new plows, drills, harvesters, and new breeds of animals. The progress was slow, but persistent; and with it American farming changed in fifty years from a matter of subsistence to an enterprise for profit.

The use of guano is particularly interesting in this regard. Maryland farmers found that they were often being cheated by adulterated supplies, and in response, the state established the post of State Chemist to verify the quality of the product. Eventually an increase in the price of the product prompted Congress to fund exploration of sources close to home and to annex the Jarvis and Baker Islands in the Pacific to assure an uninterrupted supply. Thus the use of fertilizer led one state into joining the offices of government and industry and the nation into an aggressive foreign policy to assure domestic supplies of a basic commodity. Here are the first tentative shoots of what even the partisans of progressive farming have to recognize as today's often problematic entanglements of agriculture with business and foreign policy.

Throughout his book, Danhof praises the courage it took to spread progressive farming practices across this new land. To his way of thinking there is nothing inexorable about technological progress. It requires many individuals finding toeholds for a risky step up: "It was left to men of superior foresight and energy to assume leadership and to demonstrate the profitability of operational plans that with little more capital but with substantial long-range planning would assure the maintenance and even increase of fertility" (p. 277). From the beginning, the progressive farmer kept as close an eye on his soil health as he kept on his balance sheet: "The basic characteristics of the type of farmer who provided leadership were a sensitivity to market conditions as guides to determining products and a constant reevaluation of routines of production in the light of new alternatives, followed by a willingness to make whatever adjustments appeared advantageous" (p. 280). These were the admirable farmers who broke new ground.

Progressive Farming Today

Milton M. Snodgrass and L. T. Wallace share Danhof's admiration. In the last chapter of their book *Agriculture, Economics and Resource Management,* they express dismay, in fact, that progressive farming still has its carping critics. They think this is particularly unfair in view of its astonishing accomplishments in growing more and more food with fewer and fewer farmers (freeing the rest of us for interesting work elsewhere).

It is interesting to read their book and especially its final chapter, "Persisting Problems in Agriculture," after reading Danhof's history because they discuss at length the same early resistance to modern progressive farming that Danhof talks about in passing. In one way they are more understanding than he is. They find this resistance rooted in old-fashioned virtues. Colonial farmers believed that farm owners had the right to make all the decisions about how to operate their farm, and that any government intervention in the process violated their basic right to be left alone. Thrift and hard work was all that was needed for a farmer to provide a good life for his family and the dignity of a prosperous old age for himself. These farmers fought and died for these principles in the American Revolution. They believed that they had earned the right to hold them, whatever the arguments to the contrary, and so they could make life hard for the progressives with an open conscience. The point Snodgrass and Wallace make about this history is that this old-fashioned obstinacy was in many ways appropriate to its time and place. Only hard work clearing and holding the land could have grown decent supplies of food at the beginning. Ironically it was the attitude that fit the technology and the economy of the day.

The two authors, however, share Danhof's conviction that these virtues became somewhat less sufficient as the economy and technology changed during the nineteenth century and soil lost its fertility. Then it was necessary for a hardworking farmer working the land to use the methods of progressive farming to protect the enterprise. Nowadays these virtues have become completely inappropriate even while politicians continue to extol them.

Furthermore, they believe that nowadays the concept of the family farm is "nebulous" to begin with. There are many kinds of farms run by a single family unit, from small subsistence farms in rural New England to huge grain farms in the Midwest. The term is more political than descriptive; it is used to talk about how people wished things were, instead of how things are. And therefore it is a dangerous term because it obscures the fact that no matter whose name appears on the deed of a farm, farming is an industry requiring state-of-the-art tools and an entrepreneurial state of mind. Changes in technology require changes of mind. When farm boys such as the McCormicks, the Deeres, and the Armours left the farm to become captains of industry, "they helped reshape older rural value judgments" so that "sociopolitical attitudes were modified. By the mid-twentieth century, with farmers eligible under Social Security and with burdensome surpluses at hand, opinions that farming was first a way of life and second a business have been reversed in the minds of many farm operators, particularly those with college training" (p. 498). After all, it is no longer "simply sweat and soil that produces our food, but pesticides, additives, antibiotics, and all sorts of chemical phenomena used to increase production, to improve and preserve our products" (p. 505). Consequently, Snodgrass and Wallace hope "perhaps the sacred family-farmer cow will die an evolutionary death," and that land-grant technicians will pull the plug.

It is true, they say, that many farms might remain family farms for a time if their owners use modern technology wisely enough. Most of this technology saves labor and thus could enable a small family to run a large chicken- or grain-growing operation. But the trend is against family or single farmer ownership because the new equipment is very expensive, requiring capital investments usually beyond the means of individual farmer owners. As marketing food becomes as important as growing it, fewer farmers can hope to grow and sell a product as part of a single operation. This takes more know-how than even farmers' traditional co-ops can supply. It takes a modern competitive corporate structure to manage the efficient use of resources and to find the best price for the producer at the lowest cost to the consumer. This structure works best when free from government regulation of any kind.

The two authors believe that even if the old work ethic and the family farm are values of the past, that does not mean that modern progressive farming is without a heart. It is courageous in a new way. Now the progressive farmer, whether a single operator or the member of a team, faces the challenge of learning the ever new and more sophisticated techniques today's farming requires to grow the vast amounts of food a hungry world must eat to survive and prosper. The problem in the nineteenth century of how to use the same field again and again has become the problem in the twentieth century of pulling ever greater yields from the same limited acres. The authors call the progressive farm of the present and (they hope) future that the land-grant system has wrought an "institution" with "super efficiency" for the function of feeding the nation. In other words, it is their fundamental argument that modern technology should determine what modern farming virtue is. By the same token, concepts of what farming virtue is should never be allowed to regulate the way technology is used.

The Counterargument

The counterargument is basically that virtue does not change, in agricultural matters or in any other, even if technology drastically changes the way we farm or do anything else in our daily lives. To be honest was just as hard for an early colonial, be he a cobbler or a farmer, as for a modern shoe saleswoman or an agribusiness person. To be a good farmer now, as before, requires taking good care of the land while taking care of the consumer, whether the consumer lives on the farm with the farmer or off in the cities or the suburbs, or even overseas. Farm virtue does not necessarily determine the precise size of the farm or who owns the farm. But it does condition how the farmer farms, and to the extent it does, it puts a limit on what a farmer can do to the land in addition to his or her responsibility to harvest its bounty. Anyone using this argument will usually agree that today a farmer has to know many more things than his colonial or even nineteenth-century ancestor had to know about how to use tools and machines while covering the ground; but that a modern farmer does not know anything more than a colonial farmer or an ancient farmer about what good farming is.

One could use this counterargument to reject the modern practice of highly technological farming altogether. This farming could be said to have no soul because it no longer allows the farmer to preserve the health of the soil or to deliver healthy products at a fair price for a fair profit to accessible markets. One could argue in light of what Danhof and Snodgrass and Wallace say that this state of affairs in highly ironic. If progressive farming began in this country as a means of saving the fertility of the soil, it has now developed to the point where sometimes it scatters that fertility irrecoverably in the form of wind and rill and sheet and gully erosion. If progressive farming began as a means of keeping the soil healthy, it has now developed to the point where certain soil conservation methods like ridge-tilling or no-till require spreading poisons on the ground to control weeds. New methods of pest control develop resistant strains of pests. New, highly potent chemical fertilizers leach through soils and into the watershed where they sometimes cause eutrophication. These are counterexamples used by many people who have decided in the last twenty years to return to what seems a simpler and healthier life doing subsistence farming in the outback, wherever land prices are cheap and the living looks easy. It is close to the rationale the Amish use as they hold on to preindustrial tools and even early nineteenth-century clothing.

But even the farm fundamentalist who would make this radical judgment, including the Amish farmer, would not necessarily be averse to progressive farming as such, as Danhof describes it. Such a person is very likely to use the progressive techniques of the nineteenth century, spreading manure, rotating crops and animals and fields, and even using many other techniques of organic farming and gardening that have been developed in the twentieth century and have been reported in such periodicals as *New Farm* and *Organic Gardening.* The argument such a person would have against Snodgrass and Wallace, if not against Danhof, is that nowadays much of what is called progressive farming has gone too far in one direction. It has become efficient and productive at the expense of quality. Nowadays it is just as important to think about the old-fashioned virtues of stewardship as it is to think about newer and more efficient technologies. Holding on to the traditional virtues of farming in the modern world allows one to feel for the balance point between exploitation and care, between the needs of the farmer and the consumer, farmland and marketplace.

In this chapter we give the burden of the defense of thoughtfully progressive farming to Jim Hightower, currently (1989) the secretary of agriculture of Texas. In his book *Hard Tomatoes, Hard Times,* he develops his argument beyond the fundamentalist position. He argues that modern progressive farming can be very sophisticated and mechanized, much more than the Amish will allow their farming to be, without losing the fundamental virtue of caring for the farmer and the farmland in the process. This is a more difficult argument to make than the purely fundamentalist argument; and to make it well, as Hightower does, allows him to engage head-on the partisans of progressive farming such as Danhof, Snodgrass, and Wallace. Their debate provides a good basis for making well-rounded decisions about the cases with which this chapter begins.

Hightower has biting things to say about what he thinks are the worst abuses of the modern land-grant university system. He singles out its development of a tomato that tastes like cardboard, handles like a rock, with little nutritional value, because it has been bred solely for machine harvesting and a long shelf life. Such a hard tomato is a metaphor for a system that has developed an agriculture not for the benefit of the farmer, as Wendell Berry would hope, not for consumers, as Gwyn James would hope, not even for good economy, as Theodore Schultz would hope; but solely for the benefit of profit-seeking businesspeople: "Overwhelmingly, agricultural research continues to be committed to the technological and managerial needs of the largest-scale producers and of agribusiness corporations, and it continues to omit those most in need of research assistance" (p. 25).

Hightower traces the roots of the hard tomato back to the allocation of resources the USDA has made among nine designated problem areas. The distribution, he says, is highly inequitable. A full 68 percent of its research funds support projects in the following three areas: the management of natural resources, the protection of crops from insects and diseases, and the control of production costs. These projects tend to be capital intensive and to require the use of sophisticated machines and chemicals. They are the areas of most concern to large farmers and agribusiness. In contrast, only 17 percent of the USDA budget supports projects in the following three areas: improving the quality of farm produce, increasing the efficiency of the market system, and expanding the export market and assisting developing nations. These are projects of concern to everyone involved in agriculture. Finally, only 15 percent of the budget supports the protection of consumer health, the improvement of rural living standards, and the promotion of community development through recreation, environmental protection, and increased economic opportunity. Hightower calls these three relatively neglected projects "people-oriented" and they are slighted, he believes, simply because the return on investments in them cannot be directly measured in dollars. (See chart 4.1.)

CHART 4.1   Allocation of Resources by the USDA

Capital-intensive projects

Management of natural resources

Crop protection

Cost control 68%

Marketing projects

Improving farm products

Increasing market efficiency

Expanding export opportunities 17%

People-oriented projects

Protecting consumer health

Raising rural living standards

Promoting rural community development 15%

SOURCE: Jim Hightower

Hightower believes that this lopsided allocation is even worse than it appears, because the research in capital-intensive technology works directly against the goals of the people-oriented projects, and thus over time, these latter projects will become less easy to justify. Why have a project to help alleviate rural poverty, for example, if there are no poor farmers left? If things keep going the way they are, soon all poor farmers will have been replaced, along with their middle class and upper middle class fellow farmers, by the managers that run big spreads for large corporate investors. Increasingly all farmers have two choices to make: "either get *with* the new production efficiency technology as we are developing it, or get *out* of farming business. These are not very inspiring alternatives," Hightower believes (p. 68).

From his perspective, progressive farming, however admirable its original motives might have been in the nineteenth century, has become in modern times a Frankenstein monster devouring its own children. In fact, it is worse than Dr. Frankenstein, who at least did not know what he was doing in making the monster. Today's agricultural researchers know precisely what they are doing: "The truth is that mechanization is perceived as an essential, tactical step within the land-grant community's broad strategy of making agriculture strictly efficient" (p. 32). But this is efficiency of a special sort—"again and again the message is hammered home—machines either exist or are on the way to replace farm labor" (p. 33). The ultimate goal is the vertical integration of agriculture, in which all decision-making power is taken from the individual farmer-operator (employee) in the interest of a "total food system" (p. 43).

Hightower believes this state of affairs is *not* economical in the long run, whatever the statistics to the contrary land-grant college boosters muster in support of high-tech efficiency, because the "USDA does not add on the enormous social and cultural costs that also are products of the agricultural revolution" (p. 22). These are the costs of resettling displaced farmers, replacing precious topsoil lost in harsh mechanical and chemical agriculture, and the cost of higher taxes we all have to pay to subsidize the research of large corporate interests. Like Wes Jackson and Wendell Berry, he has deep concerns about the environmental pollution and social displacements modern scientific agriculture has wrought. He has especially harsh things to say about Purdue University's Research Foundation and its School of Agriculture's use of public funds to do research on golf course grass and football fields. Basically he argues that there is no critical or adversarial aspect to the land-grant system's ag-research. It does not question what it should be doing, by and large; it simply serves the needs of industrial interests.

Instead, Hightower believes that the land-grant system should return "to the historic mission of taking the technological revolution to all who need it, rather than smugly assuming that they will be unable to keep pace. Instead of adopting the morally bankrupt posture that millions of people must 'inevitably' be squeezed out of agriculture and out of rural America, land grant colleges must turn their thoughts, energies and resources to the task of keeping people on the farm, in the small towns and out of cities. . . . In short, it means putting the research focus on people first—not as a trickle-down after thought" (p. 64). In these words Hightower shows that he is not far from Danhof, Snodgrass, and Wallace in believing that advancing technology in agriculture once was and still can be a moral activity. To a certain point advancing technology improves the productivity of agriculture and the quality of the farmer's life. Hightower differs from the other three when he insists that to preserve its moral integrity, technology should be advanced with restraint—with sensitivity to the pleasures and plights of ordinary people who love to stay close to good soil.

What-if? vs. How-to?

Danhof, Snodgrass, Wallace, and Hightower ask speculative what-if questions about agricultural technology. The three progressives worry—what if we abandon highly technical modes of farming and a mass famine results? The sustainer worries—what if technology of a certain kind ruins our soil or destroys something precious in our social system? These are broad, philosophical questions people can debate at length in the classroom or around the farmhouse fireplace in the wintertime.

But most of the questions asked about technology, at least on the farm, are practical how-to questions. A farmer wants to know how to stop erosion on his particular field and asks an extension agent about what practical steps he can take. Another wants to know how to use a system of integrated pest-management in her fields, or which tillage system will generate the greatest profit. Like most of us, farmers find themselves caught up in the middle of something that has been going on for a good long time. They want to learn as quickly as possible how to do what everyone else around them is doing. Even the scientists who discover the procedures and the engineers who concoct the tools we use are asking how-to questions all the time as they proceed in careful steps from what they already know how to do to something new that might get a job done better.

Whatever new they discover is often not all that new. In a personal correspondence with the author, Professor Larry Butler, a biochemist in Purdue University's School of Agriculture, puts it this way:

> I operate under a system in which it is assumed we are devising *innovative* solutions to traditional problems (pests, etc.). But what we actually are doing is using our chemistry to understand and appreciate the scientific basis for *traditional* solutions to these traditional problems. For example, we discover that farmers soaking sorghum grain with wet wood ash do so because the aqueous alkali detoxifies the tannin; traditional sorghum processing often includes fermentation, we find, because anerobic conditions reduce S-S bonds in the major seed proteins, making them more digestible; farmers traditionally have treated striga-infested fields with wood ash, we find, because alkali inactivates the chemical which signals the striga seed to germinate.

Speaking as an agricultural researcher, Professor Butler comes to the humble conclusion that "germane innovative solutions to farm problems are truly rare; usually the traditional agriculturalist has been there ahead of us!" In his view, the special contribution of the researcher is simply to discover how to adapt the traditional functions of farming to modern circumstances. This is the gist of whatever technical advice an extension agent will be passing to a farmer with questions.

Another way to say this is that most of the information written about the technology of agriculture is written in the user's manuals that comes with every new tool a farmer finds coming to hand. Often these manuals are complex, or the systems in which they are used are complex. It almost seems necessary for a farmer to have a college education or to have taken extensive training programs in order to use the technology correctly. Again most farmers find themselves in the middle of things and have to work very hard to become a part of what is going on.

Nonetheless, even if most of the talk about the technology of agriculture never asks the broader what-if questions directly, we can detect their presence—sometimes only in shadow—while reading even the most practical set of instructions about how to do something in a field. Take for one example an essay by E. L. Knake, "Cutting Costs of No-Till." It is a nuts and bolts how-to article addressed to farmers with a considerable amount of sophisticated expertise in the use of chemicals in farming. The author explains the value of advanced planning in preparing for different kinds of field conditions. This is a typical paragraph:

> If you are going to plant corn no-till in soybean stubble, you will not have a lot of crop residue to contend with. Using an early preplant (EPP) application such as atrazine and Bladex can help to burn down existing vegetation—smartweed for example. In addition, the triazines will give residual control. Depending on the kind and size of weeds and when you get into the field, a little Paraquat or Roundup may be necessary. You may need a good grass killer such as Lasso or Dual about the time seeds of grass weeds are ready to germinate. (p. 13)

Beneath the quasi-jargon we can detect a concept of agriculture as a mighty technological struggle against the nature that the farmer has "to contend with." The farmer "burns down" weeds, "kills" grass in an effort to "control" the life in his field. In the same essay the author uses terms from warfare—"And again, you have Paraquat and Roundup in the arsenal if needed"—and football—"all types of vegetation (living or dead) can be tackled" (p. 14). He uses the words "control" and "challenge" frequently. In the shadows behind the main purpose of giving a farmer practical advice we detect the assumption that farming is a battle of wits and weapons with a very difficult adversary. Another assumption is that the beleaguered farmer has allies in the agribusiness industry. Although this is presented as an informative essay, the author has no qualms suggesting the specific weapons the farmer needs to use by brand name. Many of the same names show up in ads in the same periodical in which he is writing. Throughout the essay there is an easy assumption that the chemicals the author cites are as much a part of farming as the food crops they protect. At one point in his essay he assures his readers that although there are certain weed problems for which there is currently no ready chemical solution, "there are several more herbicides in this category that are in the development stage." Thus, he hints, scientific farming will resolve the problems scientific farming discovers. It is not hard to determine what this author would mean by "progressive farming" and that it would not resemble what Jim Hightower would mean.

Take for a counterexample an essay by Fred Zahradnik, "Nature's No-Till." It is not intended as a direct answer to Knake and is really not the same kind of article at all. It is less how-to and more a journalistic report of scientific work being done to develop a different concept of minimum tillage farming. But we can still contrast the background language used in both articles. Here is a typical paragraph describing Wes Jackson's efforts at the Land Institute in Salina, Kansas:

Jackson dares to dream of vast, natural prairie-like fields filled with a diversity of plants that seldom need tillage, fertilizers or pesticides, yet yield grains and beans enough to feed mankind. He foresees an end to the paired problems of soil erosion and loss of soil organic matter. He and his staff are pursuing those dreams, one arduous scientific step at a time (p. 38).

Now the emphasis is on a natural farming, as diverse as nature, and hopefully as bountiful. Problems remain to be resolved in the future, the author admits, but it is a future in which the farmer will work as a help-mate of nature, not as her adversary: "Just think: A relative of the humble timothy in your hayfield may one day feed the world!" (p. 40). He describes the tastiness of a muffin made with the flour of a perennial wild grain the researchers are trying to domesticate to product higher yields. Hightower would doubtless also like the taste.

We now look at a few other important discussions about agricultural technology where what-if questions are being asked along with how-to questions.

Animal Rights

One of the more emotional debates about agricultural technology concerns the rights of animals that farmers raise for food, fiber, eggs, and other products. Animals have long been considered chattel, to be raised and slaughtered according to the needs and even the whims of the owner; and although farmers usually have a great deal of affection for their animals and treat them well—and indeed have to in order to raise them efficiently—it is something new to instruct farmers about an animal's fundamental right to a decent life as certain philosophers have in recent years. Livestock and poultry farmers are understandably suspicious of any arguments that restrict their ability to raise animals and make money as they see fit—whether the arguments come from the Federal Drug Administration or the American Philosophical Association. (See chart 4.2.)

CHART 4.2   Arguments for Animal Rights (in ascending order of respect for animal independence)

Animals have no independent rights. Humans have a right to use all animals as humans see fit.

Domestic animals have a right to lead a comfortable life until the day they are slaughtered.

All animals, wild or domestic, have a right to lead a comfortable life in accordance with their natures.

All animals have a right to a natural death.

All animals have a right to preserve their own genetic make-up and to change it only through a process of natural evolution.

The issue of farm animal rights is particularly pertinent to any discussion about technology in farming. In livestock or poultry raising, the link is short between what technology does to a food product and what that food product does to its human consumers. Sometimes the link is an emotional one. If food-raising technology makes animals feel bad by crowding, sickening, or terrorizing them, it makes some consumers feel bad about eating their flesh. That is because animals are more like human beings than other foodstuffs. They have emotional reactions to miserable conditions and pain. Like humans, they will often try to avoid conflict and life-threatening situations. Arguments about animal rights in response to human sympathy with animal suffering tend to be idealistic and philosophical. Sometimes the link is a physical one. If food-raising technology makes animals sick or loads their systems with powerful chemicals, the disease or drugs can have many of the same effects on human consumers. Animals and humans, after all, are susceptible to many of the same infections from bacteria and viruses or have similar responses to drugs. Arguments about animal rights in response to human anxieties about eating tainted meat tend to be pragmatic and concerned with health. Our brief review of this topic considers an idealistic protest in favor of vegetarianism by Peter Singer and a rational response in favor of humane slaughter by R. G. Frey. It considers a pragmatic alarm sounded by Orville Schell and a brisk economic response Schell cites by Booker T. Alford, a spokesman for a company making growth-inducing chemicals. It concludes by considering the question whether human experimenters have a right to change the basic nature of animal species by altering their genes to make them grow faster and more efficiently.

In several edited collections and in his own books, the Australian philosopher Peter Singer has argued, as the title of one of his essays puts it, "all animals are equal." By that he means that humans too are animals, and they should extend the same rights to animals of other species that they extend to other members of their own. We should do so because we share with all other animals the curse of suffering from pain, both physical and psychological. "If a being suffers, there can be no moral justification for refusing to take that suffering into consideration" (p. 154). Singer accuses any human being who does refuse of "speciesism," his word for an unethical prejudice in favor of our own species and a discrimination against animals of others. To him, farming and raising animals for food is speciesism: "our practice of rearing and killing other animals in order to eat them is a clear instance of the sacrifice of the most important interests of other beings in order to satisfy trivial interests of our own. To avoid speciesism we must stop this practice, and each of us has a moral obligation to cease supporting the practice" (p. 154).

This is an absolute argument that considers the use of *any* techniques in the raising of animals for food as immoral, because the entire enterprise is immoral. No matter how kindly a teenager participating in an 4-H competition is to the bull calf she raises for the county fair, her intent is to sell the animal to a packing house or a restaurant, with only the ribbons she wins as a keepsake. Singer is really only asking a what-if question: what if we treated animals as well as we try to treat ourselves? Some animal lovers go even further and suggest that keeping live animals captive for their milk, wool, or eggs is also immoral. It restricts their freedom for a selfish human purpose. All penned animals suffer from crowding or concentrations of pests in some way or another. Humans can do quite well eating berries, vegetables, and fruits and wearing canvas shoes.

R. G. Frey, a British philosopher, raises the question whether the vegetarianism which Singer and other animal lovers thereby espouse would not be unfair and cause pain to many animals of our own species. Many people, with good moral conscience, make their living in the meat industry, both as farmers raising animals, butchers cutting and packaging meat, and grocers selling it in retail trade. Furthermore, Singer's solution in particular might be a bad one because it is too idealistic. Western consumers are simply not going to warm to the idea of giving up meat. There are too many traditions and tastes that depend on meat—eating turkey at Thanksgiving time, for example, or lamb at Passover and Easter. If one really hopes to curtail animal suffering, one should develop new, painless techniques of meat raising for housing animals comfortably while they are being raised and for slaughtering them without pain: "That is, we are confronted with different tactics for combating the pains of food animals, and the central issue between them becomes simply the degree of effectiveness in achieving this end. The determination of which of the two tactics [vegetarianism or painless slaughter] is more effective in lessening animals' pain is not a piece of theory but a matter of fact. If technological developments succeed in the encompassing way the one tactic envisages, then it may well be, on grounds of effectiveness, the preferred one, as new and better pain-killers, administered painlessly, reach more and more animals" (p. 34).

It is not hard to understand why many livestock raisers would be suspicious of even these benevolent suggestions. They would see them as the beginning of a movement that might eventually restrict severely the ways in which they raise their animals and make their money. The new interest in farm animal rights has not developed in a vacuum. It can be seen as a response to the new technologies of high-density feed lots for cattle and confinement houses for hogs and chickens that are part of the new verticalization systems for egg and meat production. These technologies crowd animals uncomfortably as part of an economic tactic to decrease the space a farmer has to maintain to raise an animal. Hogs in confinement butt old tires or the walls of their pens or worry each other continuously. Like people and other animals, they need a certain amount of room around them to feel comfortable. Chickens in broiler factories peck at each other down the pecking order, with no escape for lower-level hens in a small wire cage. Cattle in feed lots stand in manure surrounded by clouds of stinging flies. Considering the circumstances under which many animals are raised these days, the day of slaughter becomes a day of release and blessing.

According to Orville Schell in his recent book *Modern Meat,* if many stockmen have any worries about crowding animals in these new technologies of high density, it is not likely to result from any moral concern for the well-being of animals. They are likely to have strictly business interests. Farmers (and livestock researchers) experiment with where to draw the fine line between housing animals in the minimum amount of space they need to grow normally and crowding them to the point where their growth rates become affected. The same people are likely to worry about the increased likelihood of diseases spreading quickly through the feed lot or containment pen and stunting or killing valuable livestock. Increasingly, animals are immunized by innoculation or food additives. According to the Animal Health Institute—the Washington lobby for companies manufacturing pharmaceuticals for animal use—American farmers spent more than $242 million on antibiotic feed additives in 1980. The Office of Technology Assessment, a research arm of Congress, issued a report in 1979 noting that virtually all commercially raised poultry, 70 percent of the beef cattle and 90 percent of the swine reared in this country consumed antibiotic additives as part of their daily feed regimen (pp. 19–20).

Schell makes it clear in his book that he himself has raised stock. He does not share Singer's compunctions about slaughtering animals for food. He would be sympathetic to Frey's rationale for painless slaughter. In the opening pages he describes himself giving a sow an injection to ease the difficult birth of her litter. But even as a stockman himself, he says, he has come to share the fears of some scientists that the widespread treatment of some animals with antibiotics might be encouraging the development of antibiotic-resistant bacteria. Some of these bacteria are harmful to human beings as well. The World Health Organization issued an alarming report in 1978 linking human and animal health problems:

> Outbreaks of infection due to drug-resistant organisms are an increasing problem in both the developing and developed countries. This problem has been brought into prominence by the recent widespread outbreaks of enteric disease caused by drug-resistant organisms; delayed recognition of drug resistance has, on several occasions, caused unnecessary suffering and loss of life. ... The problem is global and is the result of the widespread and indiscriminate use of antimicrobial drugs in man and animals (Schell, p. 18).

Restricting the use of antibodies in high-density animal raising, however, would now be doubly difficult since the discovery that routine sub-therapeutic doses of antibiotics help animals gain weight. It is not known precisely why. Possibly small doses of the medicine kill off bacteria in the animals' intestines that consume some of the value of the animal's feed. Now the animal can absorb more of its nutritional value directly into its own tissues. Schell finds it particularly alarming that even the companies that make and market the growth chemicals are not completely sure what they are doing. They are only sure of their profitability. He cites Booker T. Alford, who does not share his alarm. He is a technical-service consultant from the Animal Industry Department of Cyanamid's Agricultural Division:

> One theory is that these antibiotics make pigs and other animals grow faster and more efficiently by limiting subclinical disease. Another theory is that somehow the drugs increase the animals' metabolism. A third theory is that antibiotics increase the absorption rate of nutrients in the animals' guts. Frankly, we don't quite know how or why they work. But I can tell you that if you took these drugs away from the farmer, you would be costing the consumer over half a billion dollars in increased meat prices (p. 10).

What is remarkable about Alford's admission is that for his company the how-to question has become purely a question of how to make money for the company and the farmer and how to save money for the consumer. The question, How does it work? becomes trivial. The question, Could it be dangerous? is not asked.

In the face of this ominous ignorance, Schell's argument against the indiscriminate drugging of animals is less philosophical and more pragmatic—even conservative. He reasons that only to the extent that food animals are raised naturally can we be sure that they are healthy for human beings to eat. Happy pigs; wholesome chops. Schell makes a subtler point about the single family of food animals and humans in his conclusion:

> Like the air we breathe, the water we drink and the food we eat, we find that bacteria are not the exclusive province of any one country, any one part of the body, any one person or even any one species of animal. They are more like an unseen matrix connecting all forms of life. The discovery of the specific mechanisms by which resistance to antibiotics is transferred between living things reminds us that in spite of our biological diversity we are all, in the last analysis, inextricably joined (p. 326).

Where Singer gives a moral argument that animals and humans should be bound by a basic sympathy, Schell gives an ecological argument. Animals and humans occupy a single environment. If we crowd them, sicken them, force them into unnatural growth with powerful chemicals, or decrease their resistance to increasingly virulent bacteria, we eat the consequences. Whatever we do to animals we eventually do to ourselves.

In recent years the debate about animal rights has been given almost a theological basis by Jeremy Rifkin when he argues that present experiments in bioengineering threaten to take away an animal's God-given right to its own nature. He filed suit in district court in California to stop USDA-sponsored research in moving growth hormones from one mammal to another. This entailed sometimes even introducing the growth hormone of a human into that of a pig to see whether the introduced hormone might escape biochemical feedback loops in the recipient animal and thereby stimulate faster growth. If this could be done, it might be a boon to livestock producers who could raise animals to market weight much more quickly. But Rifkin argued that there should be no crossing of species barriers in mammals; and he based his suit on the failure of the USDA to file an environmental impact statement assessing these experiments before beginning them, as required by federal law. But his underlying reasons for opposing such research can be found in his book *Algeny* in which he expresses his deep dismay over the direction human bioengineering seems to be taking in both agriculture and in our whole way of looking at the world.

Bioengineering is an attempt through gene splicing and the recombining of genes to streamline the process of evolution in a purposeful way, Rifkin says, so that we can change living things to suit our preception of our need for them. He cites the molecular biologist James F. Danielli, who predicts that he and his colleagues will soon be able to speed up nature's way of doing things to the magnitude of one billion times a year (p. 220). Eventually we might be able to engineer the plants and animals we eat to match precisely the environment we want to raise them in. We shall take care of our pest and weed and disease problems at the factory as it were and make things as easy for the farmer in the field as it is for the consumer picking up cabbages in the grocery store. This might sound desirable, and it might seem like the ideal of progressive farming the early nineteenth-century progressives dreamed about when they began using their animal manures in a thrifty way.

Rifkin, however, sees it as a very dangerous trend. It would lead to a new, utilitarian view of nature where its existence seemed to serve only human needs and not its own. Perhaps worse, it might be a short step, he fears, from the idea of a perfect ear of corn or pig to the idea of a perfect human being because the same bioengineering that worked in the farmyard could be turned on the dominant mammal of the day. Even if perfection were never obtained, the possibility that it might lead to humans hating themselves and their natural world and their livestock precisely because of their failure to measure up to the new implicit standards. He wishes we would turn away from this notion of ultimate control because it will doubtless prove to be in the end another exercise in human self-deception. We can never completely control nature or human nature because we are a part of nature rather than its independent operator. Instead he wishes we could "choose participation when we could control. To acknowledge relationship when we could dominate . . . to pursue a different knowledge path, a path whose goal is to foresee how better to participate with rather than to dominate nature. To better understand the why of things as opposed to the how of things. This is the knowledge of relationship and is in strong contrast to the knowledge of usurpation that so obsesses the modern mind. . . . Two futures beckon us. We can choose to engineer the life of the planet, creating a second nature in our image, or we can choose to participate with the rest of the living kingdom. Two futures, two choices. An engineering approach to the age of biology or an ecological approach" (pp. 249; 251; 252).

An essay by Jeffrey L. Fox and Laura Tangley, "USDA Animal Research Under Fire," expresses the authors' dismay and the dismay of the research community at the nature of Rifkin's attack. They do not discuss the larger issues of Rifkin's two choices. They simply insist that American animal science is doing its job in fitting pieces of the biological puzzle together piece by piece—while looking over its shoulder at other countries doing the same research: "USDA's [Dan] Laster says, 'Gene insertion experiments within animals . . . are critical for the future progress of research. If US scientists can't do these experiments, they'll be done in other countries. It will be devastating to agriculture and to scientific and technical development. The experiments will continue until we're told to stop" (p. 7). Laster's statement bears the assumption about the inevitability of technological progress with which this chapter's discussion began. This assumption carried the day. Rifkin lost his case and several other cases since then. The experiments continue.

Conclusion

We raise the question again whether compromise is possible between these vigorously opposing positions. On the one hand, it does seem that with issues pertaining to technology in agriculture, compromise is possible. It is the nature of scientific and engineering research to be broadminded and open to the consideration of many different hypotheses. Researchers need to have various alternative explanations to test. We can see evidence of this open-mindedness to alternative techniques in two essays about minimum tillage considered above. E. L. Knake informs farmers of his best opinion on the techniques of chemical minimum tillage, while admitting that there are several other combinations of chemicals farmers can try depending on how they read the conditions in their own fields. Fred Zahradnik reports that Wes Jackson is experimenting with many different methods of what he calls a natural system of no-till. But this system might very well depend on sophisticated bioengineering of new seed hybrids for companion plantings of crops whose rates of maturity must be synchronized. In the preface to his latest book, *Altars of Unhewn Stone,* Jackson says that people in his current research group "represent the fields of genetics, ecology, entomology, plant pathology and the humanities" (p. x). Perforce at this point in their projects both

Knake and Jackson must remain open to the opinions and discoveries of other people about technical matters.

But when these researchers imply what kind of what-if questions linger in the shadows of their technical research on minimum tillage, they appear much less open-minded. In fact, they seem to rule out rigorously the borrowing of any underlying ideas about the use of technology from their neighbors across the fence. This inflexibility is probably inevitable and possibly often useful, in that it allows each one of them to develop a systematic and consistent way of thinking about agricultural technology. But it might harm the wider enterprise of agriculture if it prevents them from gaining a fully fresh perception.

Consider, for further examples, three short essays that make claims for the safety, efficiency, and even necessity of large-scale technology. They are typical of the writing that appears regularly in "in-house" publications intended to be read only by true believers in one kind of agricultural technology or another. Two essays by Jack H. Elliot and Robert S. Best that appear in the *Agrologist* make the point that the use of agricultural chemicals sometimes suffers unfairly from a bad press, as if the press were habitually in the hire of anti-technological interests. Elliot claims: "Some recent events have once again raised the suspicion that the campaign against crop protection chemicals is highly motivated and organized. Whether the activists behind the campaign are themselves victims of chemophobia, are passionately dedicated to the simple lifestyle, or just hate or mistrust industry or science is unimportant. What is important is the fact that not only the activists' views but also their misconceptions and prejudices are again being given more currency, and hence more credibility, than they deserve" (p. 6). Best takes the same line of argument but claims that in the future, because of growing public mistrust of its activities, the agricultural chemical industry is going to have to be more circumspect in pursuing its interests. Society will need to be educated to understand the meaning of "acceptable risk" in the use of agricultural technology. "No risk" agriculture is out of the question, he insists, despite a fond wish for it among the public.

Along the same lines, the editors of *Agricultural Aviation* suggest that the best way to deal with bad publicity about chemical crop dusting that might arise and threaten to curtail crop dusting in any way is to stage a demonstration "no farther than fifty miles from a major city or a state capital in order to have the best chance of attracting media and government officials" (p. 27). Perhaps this will also assure that fewer farmers or others actually concerned about chemical drift will show up and ruin the party. What is remarkable about all three writers is that none of them appear to imagine that there is anything worth listening to in any criticism of their practice; their adversaries are "against" them mindlessly. With these attitudes it would be virtually impossible to engage the editors in any meaningful debate that would lead to a productive compromise.

It is rare to find advocates broad-minded enough to admit that alternative farms—that is, owner-operated, ecologically sensitive farms—right now cannot meet all the country's or the world's food needs. At the present, only industrial agriculture can do that. But they can still insist such farming has a place. We can forget for a moment the question whether such farming can turn a profit. Often, in fact, it cannot survive unless the farmer has an off-farm income. But if it is done well, it can be a spiritually satisfying way of life for its farmers, almost a form of daily meditation on the nature of land and human endeavor. Therefore it is morally different from some other forms of farming that right now might be more economically efficient. Its survival is useful precisely as a counterexample against which to judge the benefits and the disadvantages of that other kind of farming.

Equally rare are the advocates of industrial farming who recognize that moral arguments in favor of smaller-scale or family-centered agriculture cannot be dismissed as simply bad economics. There is a certain economic value to moral values, such as the money to be made in the marketplace selling bona fide organically grown grains to health-conscious consumers. Granted, many of the arguments in favor of alternative farming are basically moral arguments in favor of a certain way of life, but that does not have to mean that they cannot withstand economic arguments to the contrary. To the extent that economic values supersede moral judgments, our culture is in danger of becoming morally bankrupt, regardless of its economic conditions.

We could all use some of the spirit manifested in an essay by Robert Rodale, the editor of *Organic Gardening* and *The New Farm*. He calls for dialogue between advocates of what is now called conventional, or industrial agriculture, and what is now called alternative, or sustainable agriculture. They do have concerns in common. For example, conservation has always been a part of conventional agriculture even as it is practiced in the United States. But in conventional agriculture it usually has been considered a form of damage control. Conservation practices correct the harm done to the soil and environment by an agriculture deemed too industrially productive and profitable to be seriously questioned. Now, according to Rodale, we need conventional conservation to marry with sustainable agriculture to become a "regenerative agriculture," as if their prosperous futures depended on one another. He concedes that "the path of marriage is not always easy. In fact, it almost always requires difficult and even wrenching adjustments from both parties. Still, there is not only an opportunity for such a union, but a need" (p. 19). When he uses the metaphor of a tempestuous marriage, he is calling for a real meeting of minds between the various parties.

We close with two essays in which writers describe their struggles to arrange such a marriage in their own minds.

In a short article in the *American Journal of Alternative Agriculture* John Vogelsberg describes himself as a retired farmer whose family has owned 800 acres in Kansas for four generations. He recalls how his father always rotated his 350 acres of cropland between animals, corn, oats, clover, alfalfa, potatoes, and watermelon. He called the method "patch farming." When John took over the farm he said he had an itch to try out the new methods of chemical farming many of his young colleagues were adopting. But he followed his father's example because "he was still looking over my shoulder," and ignored the Extension Service and farm magazines who "were pushing the idea that if you had good corn ground, you didn't really need rotations. Just put out corn every year" (p. 16). One year, he said, all the farmers in his area suffered from an outbreak of rootworms. The corn would grow half tall and then fall over because the rootworm larvae had chewed the roots off the plants. They all sprayed for rootworms, the Vogelsbergs included, and things worked well for a year. Then the worms became resistant to the spray, and most farmers in his area went to milo and wheat. "But I found that all I had to do was keep following the way my dad had done: move my cornfield and change my plantings to alfalfa or soybeans or something else. The pesticide spraying depleted our predator insects, so we had ladybugs shipped in from California by the gallons" (p. 16). He claims that as a result his corn yields remained average for Kansas, even though his production costs are lower with virtually no outlay for insecticides. Vogelsberg goes on to talk about how proud he is of the continuing health of the land his father, he himself, and now his son have farmed using a critical minimum of chemical pesticides. They have ten acres in a wildlife preserve, and eight stocked fish ponds without any of the nitrates and insecticides that pollute the nearby Blue River. "One of my biggest thrills is taking a grandson out there to fish and see that boy with an eight-pound bass on his bamboo pole" (p. 16). In this short article, he supports his practical, how-to advice on rootworm control with evidence that using both modern chemicals and old-fashioned farm smarts has been good for the pocketbook and good for the family.

There is a marvelous example of a marriage of ideas in an essay by Frederick H. Buttel, a professor of rural sociology at Cornell University. In the journal *Agriculture and Human Values* he describes his efforts to reconcile in his own mind conflicting opinions about the value of introducing for commercial use a growth hormone (bGH) developed for dairy cattle by colleagues of his at Cornell. Bovine growth hormone is a naturally occurring protein in cattle that regulates lactation. Scientists at Cornell have developed a process whereby this hormone can be manufactured in large amounts through genetically modified bacteria. Estimates predict that dairy cows fed the hormone as an additive to their feed will produce up to 30 percent more milk, which sounds wonderful. But economic impact statements suggest this will result in a corresponding 30 percent loss in the number of dairy farms through the state of New York, where the hormone was synthesized, and indeed, throughout all the dairy-producing states in the nation. Right now underproduction is not the problem facing dairy farmers. Overproduction is the problem, requiring many subsidies to pay dairy farmers the difference between the high price of production and the low price great volumes of milk fetch in the marketplace. Under these circumstances, more productive cows result in fewer cows and few farmers to meet the constant demand for milk. Many dairy farmers have become alarmed at the possibility they might be forced to become more efficient producers: "Because of the possibility that bGH will engender major dislocations in the U.S. dairy sector, dairy farmers along with public interest groups have raised concerns about this new technology at an unprecedentedly early point—three or so years before FDA approval and commercial introduction" (p. 89).

Buttel says we now have a much broader understanding of how a technological change such as the introduction of bGH affects the structure of farming. The majority of farmers do not benefit from technological change: "the farmer beneficiaries are largely limited to the early adopters—usually larger operators" (p. 92). They are able to expend quickly the capital to invest in the new technology. They reap the benefits of an aggregate supply increase and can still make a profit even as the price per unit drops. At the same time, the price drop hampers the efforts of later adopters to remain in the changing market. For this reason, smaller farms are relatively immune to the disruptions of technological change. Their off-farm income buffers them against short-term losses. But medium-size farms are vulnerable. They must produce their own sustaining income. It is often not sufficient to afford rapid capital investments. Consequently harsh markets force them into bankruptcy.

Now, one could argue that as new technology is developed, some resources may not be required in the production of the new product, and they may be released for the production of other products. So, while there is little doubt that some producers in the sector at the time of the adoption of the technology may have to go through an adjustment process and some of them will be forced out, there is no reason to believe that in the long run society is going to be worse off. If that were the case, why would we continue to invest in new technology?

But Buttel might well rebut that dairy farms form a special sector worth giving special protection from the harsh demands of short-term changes. They are mostly medium size and family owned. Given enough time to adapt to the new technologies, they can fare well in the market place. But they are particularly at risk when new biotechnologies begin to change the basic proportions of herd size, farm size, and the market price for milk. Would it not be worth preserving the social values attached to the small family farm for small periods of grace? From this perspective, Buttel says, he can readily sympathize with the average dairyman's resistance to the development of any biotechnology that will put his enterprise at risk.

But Buttel also sympathizes with the agricultural researchers who work very hard to improve the efficiency of farming: "In a dairy-oriented state like New York, Cornell University has an obligation to enable dairy farmers to reduce their per unit costs to help the dairy industry remain competitive with industries that produce other beverages and products and to benefit consumers through lower food prices" (p. 91). Researchers can also argue that the number of dairy farms in their state has been steadily declining for more than two decades. Those that remain face increasing competition from foreign producers of milk and dairy products; they need to keep their production costs low. Buttel points out that although it is true that dairy farms and herds have become fewer and bigger, they are still family owned for the most part. The American family farm in all forms survives, even as it changes its composition, because few investors would tolerate the low return on investment that farming brings. Families cultivating a way of life as well as making a living have different priorities.

Buttel concludes that the keen arguments between these two equally valid points of view on an agricultural technology are something new:

> For essentially the first time farmers are actively scrutinizing new publically-developed, output-increasing technologies at a point well prior to their commercialization. If farmers and public interest groups arrive at a common basis for opposing biotechnology research that is primarily oriented toward increased yields and output at the early research stages, public agricultural research politics will never be the same again. This—and not whether bGH leads to 10, 20, 30 percent fewer dairy farms—may well be the ultimate significance of bGH and other future biotechnologies (p. 97).

If Buttel is right, that will mean, among other things, a clear public recognition of the dependence of what-if and how-to questions on each other in any future debate about agricultural technology.

Readings

Best, Robert S. "Pesticides." *Agrologist* (Winter 1985).

Buttel, Frederick H. "Agricultural Research and Farm Structural Change: Bovine Growth Hormone and Beyond." *Agriculture and Human Values* 3, no. 4 (Fall 1986): 88–99.

Danhof, Clarence H. "Utilization of the Soil"; "Conclusion." Chaps. 10 and 11 in *Change in Agriculture: The Northern United States, 1820–1870.* Cambridge, Mass.: Harvard University Press, 1969.

Elliot, Jack H. "Agricultural Chemicals." *Agrologist* (Winter 1985).

Editors. "Fly-ins Generate Positive Press in Anti-Chemical Areas." *Agricultural Aviation* (March/April 1985).

Editors. "Milestones in ARS Research." *Agricultural Research* (November/December 1983).

Fox, Jeffrey L., and Laura Tangley. "USDA Animal Research Under Fire." *BioScience* 35, no. 1 (January 1985).

Frey, R. G. *Rights, Killing and Suffering.* Oxford: Basil Blackwell, 1983.

Hightower, Jim. "Hard Tomatoes, Hard Times: Another View of Land Grant College Research"; "Agribusiness-Agrigovernment: The Ties That Bind"; "Conclusion and Recommendations." Chaps. 2, 6, and 9 in *Hard Tomatoes, Hard Times.* Cambridge, Mass.: Schenkman Publishing Co., 1973.

Jackson, Wes. *Altars of Unhewn Stone: Science and the Earth.* San Francisco: North Point Press, 1987.

Knake, E. L. "Cutting Costs of No-Till." *Ag Consultant and Fieldman* (April 1985).

Rifkin, Jeremy, in collaboration with Nicanor Perlas. "The New Cosmic Mirror"; "Choices." Chaps. 1 and 7 in *Algeny.* New York: The Viking Press, 1983. A report on Rifkin's lawsuits appears in: Piller, Charles. "Regulating Altered Genes." *The Nation* (October 25, 1986), pp. 400–02.

Rodale, Robert. "A Proposal for Profit." *The New Farm* (March/April 1985).

Schell, Orville. *Modern Meat.* New York: Random House, 1984.

Singer, Peter. "All Animals are Equal." In *Animal Rights and Human Obligations,* edited by Tom Regan and Peter Singer. Englewood Cliffs, N.J.: Prentice-Hall, 1976.

Snodgrass, Milton M., and L. T. Wallace. "Persisting Problems in Agriculture." Chap. 26 in *Agriculture, Economics and Resource Management.* Englewood Cliffs, N.J.: Prentice-Hall, 1975.

Vogelsberg, John. "Rotations: The Key to Farm Success." *American Journal of Alternative Agriculture* 2, no. 1 (Winter 1987).

Zahradnik, Fred. "Nature's No-Till." *The New Farm* (March/April 1985).

Chapter 5

Food across the Border

The Cases

A Free Lunch

A country in Central America has been embroiled in a civil war for almost twenty years. For the last decade the United States has provided economic and military aid to its nominally democratic government as it tries to resist armed leftist armies operating mainly in the rural countryside. During this time, the government has managed to hold the insurgency more or less at bay with the foreign aid. Occasionally rebel bands cut electricity to the capital and stage terrorist attacks in the cities; but otherwise they remain in the countryside, living off the generosity of peasants who are largely sympathetic with the rebels' aims, but who, in order to make ends meet, have to sell much of their produce to government agents in charge of supplying the cities.

A devastating earthquake rocks the capital city. More than four thousand people are killed outright by falling buildings and the ensuing fires; many more are seriously injured or unaccounted for, and a very large proportion of the population is left homeless. Relief pours into the country from all over the world. The International Red Cross effectively handles the relief effort, setting up tent communities, providing medical supplies and doctors to those in need, establishing field kitchens, and in general helping the people of the city get back on their feet. The United States invokes Title II of Public Law 480, originally passed by Congress in 1954 and expanded several times since then, which authorizes the U.S. Government to ship surplus food to needy countries free of charge. The government sends huge supplies of basic foodstuffs that prevent a famine from occurring in the city, now that its links with the countryside have been completely closed off.

The U.S. State Department sends a delegation to the capital to help the country plan a long-term recovery. Congress enacts legislation providing massive aid for rebuilding the infrastructure of roads, utilities, and communication lines. The local government's agricultural minister petitions the delegation that the food aid program in particular be continued indefinitely, reasoning as follows. There is a considerable surplus of food in the United States at the present time; there is likely to be a long-term shortage in the Central American country, even after the recovery takes place, because there had been shortages even before the earthquake took place. But significantly, if the local government could be assured of adequate supplies of free food from abroad, it would be free to deal with its rural insurgency without fear of cutting off essential internal sources of supply. The delegation passes the request on to the U.S. State Department. The director of its Food for Peace office agrees to evoke Title IV of PL 480 which allows food to flow into the country on a regular basis as part of a program of economic development. There are no time limits set to the program.

The local government's plan works better than anyone had anticipated. The rural peasant economy suffers great hardships when city markets no longer buy their staple products. Large plantations in the country that grow luxury foods such as bananas, coconut, and coffee for export do much better on the global market because they hire cheaper labor when many of the peasants must apply to them for jobs. A major beer brewery develops in the capital and in a few other larger cities, processing much of the excess PL 480 grain that comes into the country into a quality product with a growing reputation on the export market. The government finds that its balance of payments stabilizes and its treasury is better able to pay back its long-term development loans to the United States and other Western nations. The insurgency fades before the booming economy.

The Price of Rice

The Portillos, a small peasant family in the Philippines, rent two hectares of land, about five acres, where they have been growing rice for many generations. The land is owned by a large landowner who lives in Manila. She charges the usual rent—one fourth of each farm's yearly yield—but she is understanding when the weather conditions are poor and the yields are low. If she decides that conditions are particularly poor, she often forgives the rent entirely for certain families so that they can use most of what they grow for their own needs. When harvests are exceptionally good, she honors the agreement to take only one-fourth of the yield. In these years, the peasants usually have a surplus to sell in the cities that allows them to buy luxury foods and appliances such as radios, refrigerators, and bicycles.

The landlady decides one day that she would like to help some of her more industrious peasants become prosperous on a more permanent basis. She selects the Portillos to receive her support while several family members attend classes run by a Peace Corps team in a nearby provincial capital. They teach the Portillos about new hybrid seeds for rice, how to intensify their irrigation practices, when and how to apply chemical fertilizers and herbicides that promise to increase their yields vastly beyond the thirty-five-sack-an-acre yield that they have enjoyed only in good years.

The Portillos worry about whether they can afford the new chemicals and seed, and whether they will be able to remember how to use them properly. The landlady agrees to help them buy the new additives for the first two years, charging them only whatever the annual rent of one quarter of the production brings and making up the difference herself. The Peace Corps team promises to provide assistance at critical times during planting and harvest.

The Portillos are hard workers. They follow directions, do what they are told, and abandon their age-old methods of manuring and irrigating for the new methods. The first year, as predicted, their yields are four times what they have been used to in a good year, and they enjoy what for them is an enormous surplus. They buy the first television set in their village with the proceeds. They discover, however, that there is a price to pay for their new prosperity. With the new chemical fertilizers the crops grow much better, but so do the weeds; and the family spends a great deal of its time working the fields pulling weeds the first year. The second year the weeds are even more luxuriant. By the third, the weeds are rank, severely cutting the yields of the year's first harvest.

This is the year when the Portillos no longer enjoy the support the landlady has supplied them in subsidizing their seed and fertilizer costs. They decide, for the first time in living memory, not to plant all the family's land to rice. They believe they can not work the weeds out of the whole plot efficiently during the time of the growing season. But they figure that with the new fertilizer, their yields for only half their land will be more than one-and-a-half times as much as they enjoyed in a good year from all their land. They are still hopeful that they will have a good year, and they do.

The following year they discover that their fertilizer bill has doubled over what it had been. They decide to change their planting plan again. This time they decide to cut back on the amount of fertilizer they use by a half. They figure that the land used to do well without any chemicals used on it at all, and they can rely on this hidden, natural help. They find, however, that the soil has lost its "spring," as one of them puts it. By using less chemical fertilizer, the land produces much less this year than it did in a bad year under the old system.

For the first time in living memory, the Portillos do not have enough food to eat for themselves off their own land, even when the landlady forgives the rent. She helps the family out by arranging for two of the males to work at a large rice plantation her cousin owns that needs workers. It is a plantation that has been successfully using the techniques of the Green Revolution for more than a decade and has remained profitable and productive because its managers have followed directions precisely. One of them is a former Peace Corps volunteer.

The Portillos are good workers. Soon two of them have responsible positions working as managers on the new plantation, and the family moves off their traditional plot and into clean and spacious housing that the plantation provides. Because of the hard work of many such families, the plantation prospers. Many similar plantations prosper, and because their labor is cheap, the Philippines discovers it can sell its rice on the world market at a price considerably below that of almost every producing nation, including the United States. Rice imports into the United States increase in volume and continue to go down in price, with the help of the high U. S. dollar. Ten years after the Portillos join the plantation, most of the farmers in the United States who used to grow rice in the Southwest have either gone out of business or switched to growing cotton or corn.

The Four-Footed Turtle

China has a history of highly productive agriculture and highly innovative farmers. But the history of twentieth-century agriculture in China has not been as fortunate. For much of the century the country has been at war. During the forty-year reign of the nationalist government, most of the country's investment went to the war effort or for industrializing the cities; the countryside suffered from a series of devastating floods and droughts.

There was some improvement during the early years of the People's Republic, established in 1949. Farms were collectivized, and although this caused much social disruption, for the next two decades China was at peace and her weather was good. The combination of a stable political structure and a benign nature, coupled with the Chinese capacity for doing hard work, resulted in adequate food supplies for the entire population. It also helped that 87 percent of the people lived on farms to begin with and could feed themselves as their first order of business. By 1957 the amount of food available per person was about 15 percent higher than it had been in 1949. Further growth seemed likely.

Mao Zedong's "Great Leap Forward" shattered these expectations, however. The object was to increase production in all sectors of the economy, agriculture included, in a very short time. But the pace was too swift and the economy broke down under the strain; agriculture suffered in addition from severe droughts and floods. Some reports indicate that as many as 30 million people might have died as a result. Recovery was slow and severely hampered by the turmoil of the Cultural Revolution during the 1960s.

In the late 1970s the new reformist government of Deng Xiaoping took control of China, and as a result of its policies, food production began to climb quickly and steadily. Between 1977 and 1984, rice production increased by 30 percent, wheat by 200 percent, oilseeds by 300, and per capita meat production by 80 percent. Government officials and foreign experts agree that these remarkable successes stem from the government adopting a *baogan* or "household responsibility system." Under this system peasants working on collective farms, once they have satisfied their production quota, can dispose of the surplus in any way they wish and can also pursue other occupations beside farming. Consequently Chinese farmers can now work for the government as well as for themselves, and the combination of communal and self-interest provides powerful personal incentives and great increases in production. As the *baogan* system spreads throughout the countryside, it goes far beyond establishing a new kind of contract between growers and buyers. Many families and individuals begin to specialize in growing a wide variety of crops on their own plots and own time, and thus the variety of foodstuffs increases along with the supply, and so does the number of restaurants and exotic foodstuffs once again able to satisfy the Chinese appetite for exquisite cuisine.

M. Williamson, an executive of an American agribusiness concern specializing in corn products, approaches Heping Zhao, the Chinese minister of agriculture. He tries to sell him a large concept. China can increase the productivity of agriculture even further with the introduction of new machines, chemicals, and research. Williamson suggests that the government do two things: one, gradually remove the official pricing structure that keeps the price of food low for the consumers in the cities but also keeps the profits of rural producers low; and two, invest heavily in Western-style machinery and chemicals that will allow farms to increase production levels at the same time. In this way new surpluses of food will rise to meet increasing demand steadily and thus stabilize the free-market price of food to the consumer. At the same time, the new technology will allow farmers to grow more food. In sum, farmers will be able to make good profits even if the unit prices of commodies go down. Williamson's firm in particular manufactures tractors and harvesting equipment and can send a team of business planners, economists, and salesmen to explain the details of the plan they have to Chinese agricultural officials.

Minister Zhao considers the proposal carefully and then decides to follow the first suggestion and to table the second for a time. He reasons that Williamson is right. It will only help agriculture to make it even freer of governmental constraint. But if China is to mechanize and industrialize its agriculture, it has to do so gradually from within its own industrial base. The social and economic upheavals of the past fifty years have taught them, they say to themselves, the value of the old Chinese adage: "The turtle will get to the pond on its own four feet."

The Issues

Crossing borders with food in mind has been going on a long time in human history. One method, invading a country and stealing its food, has perhaps the longest and sorriest history. Homer mentions it in the

Iliad and *Odyssey;* we read about the same sort of thing in many biblical books; and more recently German invaders of Soviet Russia in World War II stole not only food but boxcar-loads of topsoil brought back from Russian Georgia to the German heartland. What we are talking about in this chapter is border crossings that both countries welcome or at least allow, and that each believes can lead to an ongoing commercial and cultural relationship. (See chart 5.1.)

CHART 5.1   Types of Legal Border Crossings with Food in Mind

Two nations deciding to trade no foodstuffs between them.

Two equally developed nations trading food to their mutual benefit.

A developed nation giving away food for free or at a low cost to an undeveloped country in time of need.

A developed nation sending technology or experts to help an undeveloped nation industrialize its agriculture. Usually this is done so that both nations can then trade foodstuffs as equals.

A developed nation helping an undeveloped nation to enhance the efficiency of its traditional agriculture.

There are five basic types of legal border crossings, three of which concern us here. Let us dispose of the two that are beyond the scope of this study without elaboration. One type is for the government of one country to choose for a time to limit all agricultural exchanges whatsoever with another country. Sometimes a country might do this to punish another, as the United States did when President Carter ordered an embargo on grain sales to the Soviet Union after that country invaded Afghanistan. In this case, during the period of the embargo the shipments were limited to the quantity as specified in the long-term agreement that the United States had with the Soviet Union. The embargo caused cancellation of some of the sales beyond the minimum called for in the long-term agreement. A subtler version of this type might result from one country deciding it has no interest in the agriculture of other—perhaps because one country wisely decides another country needs to develop agricultural independence, and so refrains from trade. This first type, in all of its versions, entails more than one country simply ignoring an opportunity (or a temptation) to meddle in the farming of another. A country refrains because of a conscious decision to refrain. It pursues a policy with consequences, even if there is no obvious exchange of goods and money.

We do not consider the first type here for two reasons: there is little written about this type, quite possibly for the second reason—that no country (perhaps outside of Albania) stays away from the international market for long. Pistachios from Iran and cigars from Cuba are still to be had in the United States if the buyer knows where to look. Embargoes are usually specific, brief, and ineffective, as the U.S. grain embargo is now almost universally conceded to have been. The embargo did not hurt the Soviet Union's ability to feed itself grain and even had very little negative impact on U.S. grain sales abroad.

A second type of border crossing is to trade goods for other goods or money when the trading partners are more or less economic equals, e.g., trading U.S. grain for French wines, or Soviet caviar for Israeli oranges. Such goods are rarely bartered directly, of course. Individual import and export companies sell and buy goods at the best price they can get. Left to itself, trade in foreign foodstuffs proceeds by the law of supply and demand, affected by currency prices, shipping costs, and other business factors.

Usually this trade is not left to itself, however. Governments regulate foreign trade in various ways calculated to stimulate or discourage the exchange of certain goods across its borders. In carrying out certain governmental policies, customs officials might forbid certain foodstuffs to enter a country in an attempt to keep out pests that might be hidden in seeds or badly preserved foods, etc. Likewise, the Commerce Department might impose duties in an attempt to make the lower price of foreign goods competitive with higher-priced domestic goods.

We need to make a distinction between developed and undeveloped countries in discussing different types of border crossings. The United States imposing duties on cheap Israeli oranges is on a different order of magnitude than imposing duties on cheap Brazilian oranges. Israel is a developed country with a relatively stable economy. It could survive a tariff with some hardship, but without threat to its economic well-being. Brazil, on the other hand, is a developing country that desperately needs to earn hard currency in international trade to pay off its massive debt. To deny Brazil access to the American market could threaten to destabilize the entire country.

Thus, the plot begins to thicken when the two countries doing business are in a real sense unequally developed. The most apparently simple and straightforward type of exchange—our third—entails a wealthy country virtually giving food away to a less wealthy country, either for free, or for the transportation and processing cost, or for nominal fees, often in the soft currencies of the receiving nation. This is usually done for humanitarian reasons to help a country in need, but also because it provides a convenient way for a rich country to dispose of an agricultural surplus and to win another country's goodwill.

The fourth type of exchange entails the developed country trading with an undeveloped country in such a way as to affect substantially the way that country grows its food. This could mean encouraging an undeveloped country to grow large cash crops such as oranges, coffee, or bananas. It could mean sending an undeveloped nation technology, experts, and information about advanced techniques of growing food in order that the undeveloped nation develop an industrial-style agriculture.

The fifth type of border crossing entails one country sending technology and expertise to another in order that another country might enhance the productivity of its own native agriculture. It could mean sending an undeveloped nation technology, experts, and information about advanced techniques of growing food, but not necessarily in order that the undeveloped nation develop an industrial-style agriculture.

All five border crossings affect the way a country eats and grows its food. The last three concern us here, because with them one country deliberately changes the way another farms and eats. As with every issue in agriculture, change provokes the questions: What principle and policy should govern that change? We can roughly distinguish two principles and attendant policies.

Progressive Policy

The progressive policy pursues the principle that nations should develop cash crops for export. It favors monocultures and the use of advanced technologies for food growing on large farms. These things go together. First of all, usually only large monocultures of foodstuffs produce the volume to sustain a vigorous foreign trade. These monocultures will tend to be the crops that can be grown most efficiently in an individual country, because of its climate, soil, and the expertise of its farmers. Some examples are obvious. Oranges are more efficiently grown in the open air of a warm, wet, tropical climate like Florida's or Brazil's than they would be in an enclosed hot-house in Norway. It would probably be cheaper to grow an orange in Brazil and ship it to Norway than for a Norwegian farmer to grow and market native oranges. Better for the Norwegians, therefore, to raise salmon in large, cool-water aqua-culture ponds and trade them for warm Florida's oranges. But some examples are not so obvious. Canada and the Soviet Union have vast prairies and seasonable climates suitable for growing wheat. In recent years, weather conditions have favored the Canadians with greater harvests of cheaper wheat, which they have exported to the Soviet Union. Does this suggest as readily as the first example that one country should yield the growing of wheat to another that does it more efficiently?

This question becomes more complicated when we realize it is not only a matter of climate. One country's social and economic structure might allow it to produce a food product more cheaply than another. Perhaps its agricultural workers are willing (or are forced) to work for very small wages. When this variable is factored into the bargain, Brazil might become a more economical place to grow oranges than Florida more or less permanently, or at least as long as it can produce oranges more cheaply. Does this mean that Florida's farmers should begin looking for a crop only they can grow competitively on the world market? Without any protection against the sale of cheaper Brazilian oranges in the United States it might well be that Florida cannot even sell its own oranges to its own citizens.

In either case, of course, the law of supply and demand can be a hard law indeed. If it is allowed free play, eventually the free play of the world market will determine what is grown in every country of the world. Every one of them becomes a distinct district in a global green market.

Second of all, advocates of the progressive policy incline to believe that Western industrial-style technology should be distributed widely throughout the world. This is the technology of monocultures, after all. Every region should develop its own product the most efficiently—and thus most cheaply for sale on the world market. It is tantamount to a cheap food policy. One must be careful here. One is talking about tendencies, or a syndrome of principles. It is possible that a country such as the United States exports surpluses of grain that are not strictly speaking monocultures. Corn and soybeans, milo and oats are grown on the same fields in different years on an individual farm. Export surpluses might result each year because many individual farms at one time contribute to it. It is also possible that a country such as China might produce surpluses of rice using traditional, or preindustrial methods of cultivation on a very large scale. But by and large, the progressive syndrome encourages the development of industrial-style agriculture all over the world.

Sustenance Policy

The sustenance policy pursues the principle that every nation should feed itself first and foremost; it should be able to grow the wide variety of foodstuffs required for good nutrition and for its traditional cuisine. It favors diversified farms and the use of traditional technologies of food growing on small farms. These things go together. First of all, usually only diversified farms can produce the variety of food and fiber products that a family or a small village needs to supply its own needs. Varieties of crops and animals will tend to be well adapted to the area where they are grown by climate and environment of course, but by many other natural factors as well. These include the natural evolution of native plants and animals in the local region and the traditions of cultivation and domestication that have been practiced in the area over a long period of time. Frequently the methods used to grow indigenous crops have evolved along with the crops and the culture of the growers and consumers. A large body of local wisdom develops about how to farm continuously a certain area using local supplies of fertilizer and local methods of pest control and supplying local markets. Similarly, climate and culture will often determine what kind of cash crop small farms will tend to grow for sale in neighboring regions or nearby cities: cabbage for sauerkraut in northern Europe and olives for oil in Sicily. The laws of supply and demand will be local laws.

Second of all, proponents of the sustenance policy incline to believe that Western industrial-style technology should be introduced with caution. It is important not to disturb traditional relations between farmers and land, culture and farmers, traditional and innovative techniques. Continuity is a primary value. In fact, Western industrial-style agriculture might have techniques to learn from long-term, sustainable agricultural systems.

However, as the following discussions show, honoring these principles and pursuing these policies is no simple matter. The lines between them are not always clearly drawn. As we see today in Africa, native peoples often petition for industrial agriculture to change their way of life. Sometimes learned Western experts honor and seek to preserve native wisdom. It might be the case today that throughout the world, all five types of agricultural exchange are so intertwined and the principles underlying them all so convoluted, that the entire global market is beyond any one person or government's comprehension as well as control. World agriculture is more than a fact; it is our fate. We are as dependent on its vagaries as we are on those of the weather. By learning about how food crosses borders, we learn something about what it means to be human. We learn the best being done to try to control our fate.

We begin reviewing a number of opinions about the value and effectiveness of PL 480 in distributing gifts or grants of surplus food from the United States to needy peoples in the world. It is an example of the third type of border crossing, in many ways the best intended. The questions it raises are fundamental. Because its advocates often openly boast about its effects, the reasons given to justify or criticize it will clarify some of the less obvious reasons behind the fourth and fifth forms of intervention and trade.

The Review

The Third Type of Exchange: Free Food—Free Lunch?

The United States Congress passed Public Law 480 in 1954 mandating the distribution of surplus American food to needy countries if the surplus could not be sold in the domestic or foreign markets. Part of the original intent of the bill was to give the government another weapon in the cold war. The United States could win the goodwill of needy peoples throughout the world giving of its bounty and thus cover any potential seedbeds of social unrest that communists might otherwise cultivate.

There were originally four provisions in the act. One permitted the sale of American food in local currencies, usually soft currencies that have virtually no value in any other country except the country that issues the money. This money usually remains in the paying country and is used to support American interests there, such as maintaining an embassy or providing scholarships to American students. A second provision allowed distribution of food on a one-time basis free of charge to countries facing famine, as well as distribution of food packages on a regular basis through established relief agencies such as CARE to refugees and orphanages and the like around the world. A third provision regulated an international barter of American food for strategic goods such as essential raw materials. A fourth provision was like the second, in that it allowed the United States to distribute food freely, but now as part of an extended program of economic development in an undeveloped country. About one half of the food distributed under the act is distributed under this provision. (See chart 5.2.)

CHART 5.2   Public Law 480 as Originally Passed

Title One

Permits the sale of American food abroad in soft currency. Usually the American government spends the money in the receiving country.

Title Two

Permits distribution of food for free on a one-time basis to a needy country free of charge.

Title Three

Regulates the trade of American foodstuffs for strategic materials.

Title Four

Permits distribution of food for free as part of an extended program of economic assistance.

Dr. Don Paarlberg of Purdue University played a key role in designing the program and administered it for several years as a special assistant to President Eisenhower. Former Senator George McGovern of South Dakota was the first full-time director of the U.S. Food for Peace Office that President John F. Kennedy set up in the first days of his administration. The office was to oversee and coordinate all the services mandated under PL 480. McGovern was, as would be expected, a strong advocate of the law and Kennedy's program, describing them as "an ingenious combination of self-interest and idealism" (p. 17). As he tells the story, it was in the self-interest of the United States to dispose of its farm surpluses, which first became a serious problem in this country in the middle 1950s. The mechanization of American agriculture had been gathering steam since World War II, when it had been necessary to produce greater amounts of food to feed our armies and allies with fewer farmers, as many of their sons went off to war. The Marshall Plan of the 1940s had taken up most of the increasing productivity of American farms for distribution to war-torn Europe, as did the Korean War in the early 1950s. But with the beginning of the cold war peace, America found itself burdened with an embarrassment of riches in an agriculture that could produce far more food than the nation needed or that it could sell abroad at the time—or indeed that it could even store very easily. Yet there was no easy way to pull back on the accelerator. At the same time there were many countries of the world without the intellectual resources and social infrastructure of Europe and Japan. They could not rebuild or even build their societies to provide their citizens with a decent standard of living. Many of their people were, in fact, starving. McGovern describes PL 480 as the most popular piece of legislation ever passed by Congress. It addressed both these problems simultaneously and generously, to applause at home and abroad. It was an immediate success. By the time McGovern took over the Food for Peace Office in 1960, six years after the law was passed, one out of every three acres of American wheat was being harvested for distribution by this program.

The picture McGovern paints of rural life in less highly developed countries is sometimes condescending and perhaps simplistic when he says; "Four out of five inhabitants of the planet live in rural areas. Most of them are clinging to life by means of primitive farming techniques little changed since the days of Moses" (pp. xv-xvi). There is little acknowledgment here that primitive farming techniques might mean highly developed techniques, in the way Theodore Schultz describes them, as very finely tuned to the environment and social circumstances in which they are used. McGovern does not address the issue that starvation in a country might be more the result of wrong-headed governmental policies than the result of unproductive farming practices.

But he does declare that it was the policy of his office "not to dissipate the incentive of other people to increase their own food production in becoming overly dependent on outside assistance" (p. 6). Thereby he acknowledges that, as he sees it, the Food for Peace program is a stop-gap measure, intended to build up the strength of a country to the point where it could produce enough food to feed itself. It is easy to understand why a politician would believe such a program was good politics, and why anyone could acknowledge that feeding hungry people, as this law has enabled, is a good thing to do, whatever the motive for doing it, or even whatever the eventual detrimental effect the free food might have. This is American altruism at its best. McGovern, writing in the early 1960s, cites as an example "our effort in Algeria, where four million people were uprooted by the long war of independence. American wheat, milk and oil are now being used to pay the wages of Algerians reconstructing their country" (p. 19).

It is interesting to read Armelle Braun's essay "Rethinking Agricultural Development" published twenty years after McGovern's book in *Ceres,* the United Nations agricultural periodical. He argues that in the years after her successful revolution of independence from France, Algeria erred in relying on food aid from the United States in order to be able to invest most of its resources in building up the country's manufacturing industries. Food production actually fell for wine, dates, citrus, and summer grains, while it remained the same or increased only slightly for other products such as winter cereals, milk, meat, fruit, and nuts. Only potatoes were produced in proportions in excess of the increase in the population. The emphasis on industry was not only financial. Industrial development seemed a glamorous enterprise, the means by which the country would pull itself up by its bootstraps, and so it attracted the committment of the "youngest and best trained" college-educated citizens. The agricultural sector did not share in this glamour because national production levels, coupled with foreign aid food supplies, assured the country that there would be no shortages. The effect of PL 480 on the Algerian farm economy seems to be one McGovern warns against. It prevented a country from becoming independent in food production. The difficulty in avoiding this problem seems to be that in the initial stages of aid giving it is virtually impossible to determine what the long-term effects might be.

Theodore Schultz had already pointed out the likelihood of something like this happening in an essay published in 1960, several years before McGovern's book. Schultz argues that distributing food surpluses abroad encourages just the sort of mistake Braun complains about. It drives down the value of indigenously produced food and encourages national leaders to pin their hopes on industrial development once basic food supplies have been insured, especially to the cities. His language is strong: "What we do know is that U.S. policy concerning the use of P.L. 480 loans and grants is set against agriculture in the receiving countries" (p. 220). Schultz is careful in his criticism because he recognizes the deep satisfaction many Americans feel in thinking PL 480 is helping the poor and the hungry; and he admits that there is good evidence that bettering the diet in a poor country encourages long-term development in a basic way. Better-fed students learn more and will be able to contribute more to their own nation's economic and social growth. But he speculates that most poor countries would rather receive unrestricted grants of money to be spent as they wished, rather than supplies of food. As it costs more than half the value of every food commodity the U.S. sends abroad for shipping and processing, the United States could keep the food, give half its worth in direct, unrestricted foreign aid, and make both parties to the deal happier with the result. Receiver countries could use the money, if they wished, to buy cheap supplies of food on the open world market; or they could use the money to import the sort of capital goods that would allow them to develop their own industrial agriculture. Schultz is obviously talking about long-term distributions of food, not short-term famine relief under the second, original provision.

The Fourth Type of Exchange: Developing Trade by Developing Partners

According to Nicholas D. Kristof, writing in *The New York Times,* nations in the third world in the 1980s have now mostly turned away from the development policy that Algeria and other nations followed in the previous two decades. Their experience had been that rapid industrialization lured farmers from the countryside so that national food production fell. Governments traditionally depressed the price of even reduced stocks to insure that food would be cheap for its industrial workers. At the same time, the factories in which they came to work were not able to manufacture and sell enough goods abroad for their governments to afford to import food to make up the difference between national food production and need. Their economies stagnated, and poverty and hunger increased. Now their governments are much more likely to be investing in agribusiness than in heavy industry: "Farmers, not industrial tycoons, are seen now as the pivotal figures who can help pull their countries out of the mire of indigence" (p. 1). The shift of priorities has already had measurable results: "With increased resources devoted to agriculture in most of the third world and Eastern Europe, more countries are able to feed themselves, or even to export their agricultural products, such as grain. This cuts into traditional markets of American farmers, who are already competing in world markets with Indian peasants from the fertile Punjab" (p. 8).

By peasants Kristof does not mean traditional farmers working small plots. He means workers in industrialized agribusiness that governments and outside investors have increasingly come to capitalize on. The following writers in our review describe the strategies they think will make these workers and their new larger farms ever more productive.

Like Theodore Schultz, Milton Snodgrass and L. T. Wallace favor the worldwide free trade of foodstuffs among more or less equal partners. Usually this means between nations that are more or less equally developed. But one nation like the United States could also take active steps to create international equals where they do not yet exist. The two authors believe that it should be U.S. policy to distribute food under PL 480 always with the goal in mind of helping another country develop both its manufacturing and its agricultural technology to the point where it can trade its goods on the world market: "their rapid economic growth would improve the prosperity of the world as a whole" (p. 494). They cherish the venerable idea that international trade binds the world together into a community of common interests. The American citizen who eats a Central American banana might grow, process, or export the wheat the Russians need to create a more nourishing and satisfying diet for themselves, and so on.

Snodgrass and Wallace advocate a "modified free trade" policy governing this exchange. The United States should protect very few of its domestic agricultural markets with tariffs, quotas, or embargoes. A tariff adds a tax to goods coming from another country to raise their prices to a rough equality with domestically produced goods; a quota restricts the amount of goods coming in, consequently raising demand and price; an embargo forbids importing altogether. There are times, short times they say, when domestic production should be protected; but in the long run domestic production should compete directly with foreign production so that eventually each country will produce only those products it can produce most economically and efficiently. In the long run this would help American farmers. It would encourage other nations to drop their protections against the products American farmers can grow best.

Raymond J. Doll, Glenn H. Miller, Jr., and Richard D. Rees believe that a long-term policy like this will encourage each country to grow the crop best suited for its climate and soil. In a report they prepare for the Federal Reserve Bank, they encourage what they call "regional specialization": it insures that "resources may be more productive in one geographic environment than in another. Trade is necessary to take advantage of the potential gains from this increased productivity. If trade is restricted to protect less efficient use of resources within a nation, there is a subsidy to the inefficient resources just as certainly as if direct payments, price supports, or Government purchases at above-market prices are used" (p. 43). They make several assumptions with this idea that not all economists would share. One is that economics is the best measurement of success. If California can grow a cheaper carrot in an agribusiness field, then, strictly speaking, if they are equal in nutritional value, that is a better carrot than the more expensive one you might buy at a roadside stand in Connecticut (or Chile)—regardless of how fresh and tasty the produce.

This large-scale, abstract monetary view overlooks millions of local details and cultural idiosyncrasies to highlight what its proponents consider essential. This is not necessarily an insensitive vision, however. Fred H. Sanderson argues that people in undeveloped countries who are living "on the margins of subsistence" are living in circumstances that are "extremely unsatisfactory" (p. 1). He hopes to raise their standard of living, and in the long run this is best done by providing their culture's agriculture with the tools and techniques it needs to be able to compete on the world market. It is something akin to one baseball team lending bats and balls to a visiting team whose equipment was lost in transit. Such a policy helps the developed country in the long run as well. It encourages that country's farmers to produce only those foods for which there is a genuine and continuing need on the domestic and foreign market that can be calculated in advance. In this way the United States might avoid producing the huge surpluses that result when farmers are under the impression that the foreign market for their grain is unlimited.

Thomas T. Poleman corroborates this view by insisting that we have to abandon our notion of the noble peasant. Nobody really wants to remain a peasant or even a farmer in this world of rising expectations. Developed countries have an obligation to spread the economic and industrial revolution to the remotest fields of the undeveloped countries. Part of this obligation is moral. Poleman mentions Malthus's prediction that as human populations increase geometrically and food production increases arithmetically, eventually many people will starve who cannot find enough food to eat. Only the Green Revolution has intervened between Malthus's prediction and its realization; and now that it has, the revolution must continue if starvation is to be kept at bay. Peasants must be equipped to farm scientifically. If that brings about changes in their culture, that is all to the good. Generations of peasants have sought a better life for themselves by migrating to the West in the last two hundred years. They left poverty behind to help develop new, well-fed, and prosperous cultures. Something like this can be done for the heirs of their former neighbors they left behind. This is to say that Poleman emphasizes the population side of the food/population equation. It is population pressure that has brought about the need for much more intensive use of land. There is no other way that mass starvation could have been avoided.

Writing in *Agriculture and Human Values,* Norman E. Borlaug, a consultant with the International Maize and Wheat Improvement Center in Mexico City, urges all concerned to act fast to help them. Right now hundreds of millions of poverty-stricken people do not have enough food to eat. "It is my belief that all who are born into the world have the moral right to the basic ingredients for a decent, human life" (p. 14). His agenda is first to provide people alive today with emergency stocks of food to eat, then to educate them about the need to control the "population monster." Until these basic problems have been addressed, any thinking about the environmental matters related to farming is an inappropriate luxury.

Borlaug's convictions stem from four decades of work with the political leaders of food-deficit countries of Africa, Asia, and Latin America to raise their food production. He praises the Green Revolution in which he took part: "Certainly, the greatest satisfaction of my scientific career has been to see India and Pakistan utilize the package of high-yield wheat technologies—what we, at least in a small part, halfway around the world in Mexico helped to develop—and become self-sufficient in cereal production" (p. 9). He insists the revolution must continue. World food production must double in the next forty years to feed the hungry people as yet unborn. Foreign governments must continue to support their agriculture. Advanced countries must continue to help them. He says in his experience he has learned that these programs depend absolutely on the committment of individual applied scientists in the field. It takes a scientist from an advanced country about ten years to develop the multiple skills needed to act as an "integrator," that is, someone who can design the technology suitable for a particular locale, train local scientists and farmers to use it, and handle the politics of implementation. It takes native scientists five to eight years to develop the skills and confidence to sustain the work the integrator has begun. The task of developing such a corps is daunting. "Africa's need for trained scientific manpower for agriculture, as I visualize it, cannot be met in less than twenty-five years" (p. 12).

Bruce F. Johnston and Peter Kilby take a broad view of the problems such field experts will face. For one thing, native farmers often have a stubborn pride in their efficient but relatively unproductive methods. They already know a great deal about the best seeds to use and how to take advantage of local conditions such as climate, rainfall, availability of manure, etc. Peasant farming usually produces more than the farmer and his family need for their own strict subsistence. It gives them a variety of foods to eat and usually produces a surplus that can be traded for goods elsewhere. But the surpluses are small: "While the farming regimes of traditional agriculture are in most instances well adapted to their environment, it remains true that the underlying technology everywhere provides but modest returns to human effort" (p. 25).

The authors make a special point of the problems created by the close relationships among the farmers living in a village or an area. Because none of the farmers can count on producing great surpluses in any given year or of being able to store a surplus should one occur, all of them become very dependent on their neighbors for help in times of meager harvests and hard luck: "Old age, sickness, crop failure and injury to livestock represent the principal causes of loss of income. Because of the lack of scientific knowledge these risks are greater in traditional agriculture than in developed economies" (p. 27). Peasant agricultural societies are consequently "leveling" societies where everyone is expected to remain on more or less the same economic level; anyone who becomes more prosperous threatens to become someone who does not depend on the community, and upon whom the community will not be able to depend.

Like Schultz, Johnson and Kilby believe that to change this sort of farming so that it produces surpluses for sale on the world market requires changing *everything* about it. That includes in particular the dependence of an individual farm family on its own labor for its subsistence and its dependence on other families living in the same place. They would break up the neighborhood. Only by making them independent operators are farmers freed from the constraints of self-sufficiency for genuine and vigorous competition with their fellow farmers nearby and abroad. As the authors put it coolly: "As the security basis of reciprocity weakens, the incentives for ignoring its commands grow stronger" (p. 52).

They put their hopes for producing this change in a system of specialization, Wendell Berry's bugbear. In place of a population of mostly farmers, their children become doctors, lawyers, scientists, and manufacturers. These new professionals provide many of the services the peasant society used to supply crudely for itself—from witchdoctoring to security—as well as new goods such as detergents, electricity, synthetic textiles, plastics, etc., that substitute man-made for naturally made products. Thus, social specialization frees farming to become solely the production of basic commodities or raw materials. Farming, in turn, uses specialized tools and methods to grow its products. After a certain point, this process perpetuates itself: "The efficiency with which markets operate at any given point constitutes in itself a significant stimulant . . . to further structural transformation" (p. 42). Government helps by collecting taxes to be used to fund technological development, education, and the construction of roads, utilities, ports, and other elements in the national infrastructure, and government oversees perfecting the working of markets.

What all citizens stand to gain from this transformation of structure, the peasants included, is the opportunity to raise their material standard of living and to bequeath to their children "heretofore unimagined levels" of wealth (p. 54). The trade-off is simple: traditional communities for independently prosperous individuals.

This point of view seems impossible to refute on its own terms. Its logic seems to be the logic of history, of money, and even of common sense. Everybody wants to be as happy and comfortable as the average American (don't they?), and the average American wants to stay that way. So let us design an exchange of expertise and specialized foodstuffs that will spread the common wealth.

When we think back to Schultz's book, however, we remember one of his cardinal points was that peasant or traditional agriculture was often highly efficient. It produced all that the local population needed, along with a small surplus for cash; and most significantly, it did so year after year. It was fully integrated with its people and place. And so when Sanderson finds what he calls "subsistence . . . unsatisfactory," we have to understand that he means from *his* point of view, not that of traditional farmers. What the fourth type of intervention really entails is changing another people's way of life, and its basic assumption has to be that our way of life in the most highly developed country of all has to be a better way—even if the noble peasant is not aware of it yet. Robert Paarlberg reveals this assumption in a brief phrase in his essay on third-world development. He says he hopes rapid development "will trigger a dietary transition" in undeveloped countries (p. 175). He means that traditional peoples will gradually lose interest in their traditional foods and grow hungry for the kind of food that mass production alone can grow. He goes on to say that this transition "will increase commercial demand for precisely those sorts of farm products—especially coarse grains and feed stuffs—which U.S. producers are best suited to provide" (p. 175). Now he makes it clear that the country which will gain the most from the rapid technological development of world agriculture will be the nation that is already on the top of the grain heap and planning to stay there—the United States.

Another point of view on this business comes from reading an editorial in *Ceres,* "The Access Route for Peasants: Still Some Roadblocks." The editorial says, among other things, that "most countries suffered a sharp decline in the amount of arable land between 1970 and 1981, due to both demographic pressure and the rising power of rich and middle peasants" (p. 25). In spelling out what this means, the editor counts up some of the hidden social costs of industrializing agriculture in the third world. They are similar to those in our own country. In both, fewer farmers come to own more land, as industrialized agriculture favors the economics and techniques of large scale. But in the undeveloped countries, instead of many peasants and farmers becoming uniformly more middle class on or off the farm, as they tend to in developed countries, a few peasants tend to become rich landowners. Most of their old neighbors leave the land for dead-end lives in the city:

> In Latin America [in particular] the rapid transformation of the latifundia [big farms] into capital-intensive commercial-export farms has forced many tenants off the land, to be replaced by seasonal day labour. The worsening plight of the Latin American rural poor is reflected in the high rate of migration to the cities, the declining proportion of wages in farm costs (from 31 percent in 1960 to 18.1 percent in 1980), the decline in real wage levels, and an estimated rural unemployment rate of more than 40 percent (p. 24).

These statistics suggest that structural transformation does not take place in the idealized, progressive time of posters and slogans but rather in the real time where certain individuals, as always, find new opportunities to exploit change to their personal advantage.

Counteracting this tendency to selfishness has been hard. Several countries such as Bangladesh and Sri Lanka, for example, have tried to set up extension services to small farmers still on their land to help them farm more efficiently. But because there are many small farmers it has been impossible to find enough qualified agents to help them. In contrast, the few larger farmers are able to afford to hire all the expert help they need, and the gap between the productivity of smaller and larger farmers continues to grow apace. Virtually everything in the lives of smaller farmers conspires to make it so. Small farmers tend to live in remote areas hard to reach with new educational opportunities. They tend to be less educated and therefore less able to learn new and sophisticated techniques. They tend to be less energetic and ambitious because of their diet and traditions and thus less open to new ideas. The problems of helping them cope with rapid transformations of their agriculture are great; the editors cite an ominous warning of Erik Eckholm:

> If current demographic and economic trends are allowed to continue, one billion or more rural residents of the Third World will lack secure access to farmland as humanity enters the twenty-first century. Many of the landless will turn up in the over-flowing slums of Third World cities; some will appear as illegal aliens in the cities of the richer countries. The malnutrition, illiteracy, poor health and general powerlessness of those who stay behind will receive frequent comments in UN reports and the global media, while the sporadic violence and more systematic political activism their living conditions spawn will be described as "worrisome instability" by leaders in the world's capital cities. One way or another, the landless will be heard. (p. 28)

The Fifth Type of Exchange: Nurturing Native Agriculture

To deepen the debate we turn to some other ideas in support of the fifth and last type of foreign intervention. We consider some writers who believe that educating farmers means first of all educating them to appreciate their own values, culture, and terrain. In other words, according to this type, our modern experts would study how to make traditional agriculture more productive of nourishing food without going so far as to change its basic ability to sustain itself and to sustain its farmer's culture. Two assumptions power this idea. One is that traditional knowledge is different from scientific knowledge—but just as good.

For instance, Richard B. Norgaard does not share Sanderson's view that the practice of traditional agriculture is "extremely unsatisfactory." Instead, in an article in the *American Journal of Agricultural Economics,* he sees the practice as tailoring the needs of people to the possibilities of their particular environment—no mean task. "Traditional knowledge is location specific. It develops in the context of a unique coevolution between specific social and ecological systems. The uniqueness indicates that most traditional knowledge will not transfer directly to other contexts" (p. 876). Theodore Schultz has said as much; but unlike Schultz, Norgaard believes that this close reciprocity between land and farmer, farmer and culture, is worth keeping and even worth protecting against the aggressive competition of industrial farming that it cannot withstand on its own. His reason basically is that this sort of farming has gathered a vast amount of information about what he calls "agroecological" farming, and that this sort of knowledge is becoming more and more essential for conventional farming even in highly indus-trialized countries. After he says that "most traditional knowledge will not transfer directly to other contexts," he draws the conclusion that therefore "new institutions" need to be set up that will "capture the potential of traditional knowledge" to make it available to scientists trained in modern industrial science and technology. In this regard, re-call Professor Butler's description of his research above. He seeks to ex-plain what many traditional farmers already know intuitively.

Norgaard makes a more general point that people's basic attitudes to-ward knowledge shape appropriate institutions for collecting, developing, and disseminating information. The Western assumption that scientific knowledge is simple, general, and divisible favors centralized, standard-ized bureaucratic institutions, whereas the traditional assumption that lore is complex, specific, and interrelated favors decentralized, local, and equitable institutions. Thus, he says, "taking advantage of both objective and traditional institutions would entail maintaining institutions based on conflicting epistemologies" (p. 877). Epistemology refers to different styles of learning. Norgaard would probably have in mind the style used at institutions such as Wes Jackson's Land Institute. But he calls for more than creating new institutions. We all must work to preserve the tradi-tional and the industrial in order that both might be able to communi-cate with each other. He ends his provocative paper with a paradox. He says that these two different kinds of knowledge will always remain in-compatible with each other and will continue to develop very different technologies and social structures, but that "the evolutionary perspec-tive" requires that we "at least provide hope that we may eventually enjoy the best of both realms of knowledge" sometime in the future (p. 888).

Randolph Barker writes in a review of Norgaard's paper in the same issue that he basically shares his hope that such a marriage of views might be arranged. But he finds the obstacles to it virtually insurmount-able. The best scientific minds do not yet think this way; and right now the basic research methodologies followed in both developed countries and undeveloped countries do not allow it. Consequently there are few scientists or engineers who will be capable of understanding the need for preserving and developing native knowledge; and few to speak out in defense of the integrity of an agriculture that they cannot understand. This suggests that Western technology is indeed biased towards the in-dustrial culture that nurtures it and can only serve as a dumb tool in the hands of those who advance its cause everywhere throughout the world. Barker reasserts the basic belief in the West that the technological imper-ative is irresistible in agriculture or anywhere else.

A second assumption powering this idea is that traditional agriculture is *better* than industrial agriculture because it is the only agriculture practiced in the world today that makes ecological sense. It fits with its locale, and as even Theodore Schultz would admit, it can sustain its modest yields indefinitely in that locale. According to this way of thinking, every region should produce whatever it can produce best—as advocates of the fourth type of foreign intervention believe—but now "best" means ecologically suitable. We can assume that traditional farmers have already found the best way to grow a certain crop in a certain locale. We also need to assume that they have never been tempted to mine the soil, as the ancient sheep herders in many areas of the ancient world once did. Helping good, ecologically aware, local farmers to enhance their native productivity should be the basis of what George Cox and Michael Atkins consider an ideal "International Agricultural Policy." Such a policy should be concerned with helping other countries to grow and buy good supplies of nourishing food; but it should also secure "(1) conservation of exhaustible energy resources in other sectors of the economy, which would preserve these resources for critical roles in agriculture; (2) reduction in the dependence of agriculture production upon the most limited exhaustible sources of energy; and (3) development of flow resources for more extensive use in agriculture" (p. 692). "Flow resources" is their interesting term for solar or renewable forms of energy production from wind, water, and sun. The basic problem with industrialized agriculture in general, and the Green Revolution in particular, is that both depend absolutely on unlimited supplies of nonrenewable energy sources. The basic virtue of traditional agriculture is that it depends absolutely on limitless supplies of renewable energy sources. The marriage they hope might be possible between them would be highly sophisticated and technical, but it would also be highly tailored and specific:

> closer adjustment of machine power to job requirements; increased sharing of specialized equipment; replacement of machine use by other techniques; a shift from chemical control to increased reliance on genetic resistance and biological control; a reversion to extensive grazing instead of intensive feeding for livestock production; and the substitution of skilled or unskilled labor for operations that now consume large quantities of exhaustible energy resources. (p. 694)

Note that this hybrid technology requires a hybrid social system combining features of traditional and industrial agricultural cultures. Farmers share equipment, and they leave land unplowed for grazing. They work in teams doing work that large machines can do but without expending inappropriate amounts of unrenewable energy. As the authors put it: "In developing countries, where rural labor supplies are relatively large and inexpensive . . . the goal must be to utilize the labor force without displacing it" (p. 695).

In basic agreement with this point of view, George McRobie (a disciple of E. F. Schumacher) argues for the use of "intermediate technology" in third-world countries. He looks for tools whose costs strike a balance "somewhere between the insignificant cost of the traditional tool and the expensive Western technology; exactly where it would lie would depend on local circumstances and would bear some relation to the average income per capita in the country" (p. 33).

It is highly interesting to read in this regard an account of the North American scientist K. Briggs describing the problems he had in trying to train traditional farmers in Kenya to use industrial farming techniques for growing wheat. By training and inclination Briggs is inclined to the fourth type of agricultural intervention abroad. He believes that advanced countries should help less advanced countries industrialize their food production. But he is sensitive to the need to avoid becoming too aggressive. It is a common problem, he says, that a highly trained and technically competent Westerner arrives in an undeveloped country "fresh from his own . . . incentive society" (p. 32) with a contract to work two years toward achieving some specific goal. He or she tends to take over a project, using native researchers or farmers as assistants, and leaves with the job done but without having passed on to local people the expertise and confidence they would need to carry on such work themselves.

Briggs believes they do need to carry on such work. He himself is a Canadian, trained to work in developing Canadian wheat production and therefore not trained to tailor development programs for a country like Kenya with only a fraction of Canada's tillable acres, with a far greater variety of "micro climates" and with a very different tradition of agriculture. He describes this tradition succinctly:

An average Kenyan family with a 4 acre "shamba" (small farm) in the highlands can easily expect to produce enough maize and beans to feed themselves, with plenty left over to sell for cash. By contrast, the cash profits from 4 acres of wheat would likely not even feed the family for one year, and the wheat would not be desired for food on the farm where maize and beans are the stable diet (p. 28).

He believes that this tradition has to evolve in order for Kenya to be able to feed her growing city population, and that the local government itself will need to work out conflicts between many different policies in pursuing this goal. Currently government policies actively promote keeping as many people as possible on the land in small shambas, while at the same time asking for help from abroad in developing the large-scale grain production that requires large farms and heavy machinery. His version of hybrid development joins local people and foreign experts in tailoring a new agriculture to suit the country in which it is done.

Frances Moore Lappé and Joseph Collins frankly admit that their program "food self-reliance" is intended to keep the people in less developed countries fed as well as free. Their program has seven fundamentals. (See chart 5.3.) The first requires giving local people control over their own agricultural resources; in this way any increases in production will help them "instead of local and foreign elites" (p. 373).

The second requires stimulating large numbers of farmers to improve their own situation themselves, without waiting for directives from their own government or abroad. This will help free people from "dependence on authorities" (p. 375). In describing the third fundamental, they say that "with food self-reliance, trade becomes an organic outgrowth of development, not the fragile hinge on which survival hangs" (p. 376); it is putting self-sufficiency before international dependency without arguing against foreign trade in itself. In describing the fourth fundamental, they say that developments in agriculture should go hand in hand with education about nutrition. Traditional peoples have combined much of the food they have grown and hunted into well-balanced diets developed over a long period of time; now if food practices change, care must be taken to make sure that new combinations of food will supply the same necessary proteins and calories in the proper proportions. The fifth fundamental grows out of the fourth. "Food self-reliance makes agriculture an end, not a means" (p. 379); this means that agriculture has always been and should remain a basic way of life for the farmers and their families who live on their farms. It should no longer be accepted as a given that people are necessarily better off living and working in cities and industries. The sixth fundamental is that industry should serve agriculture instead of, as it happens in the Western world, industry creating agriculture in its own image. This means basically designing tools and systems to cultivate and enhance the productivity of traditional farms. Finally, according to their seventh fundamental, food self-reliance requires highly coordinated social planning, for which they look to China and Cuba for examples. Both these countries are now beginning to capitalize and industrialize their agriculture and to allow for much more self-reliance in their farmers. But they still are attempting to integrate agricultural production with other social and industrial goals. These two models are not the only ones to follow, or perhaps even the best, the authors admit. But they insist on the need to link economic and technological change in agriculture to the social needs of the people who grow the food and eat it. In their view, the free market is never free and never a market. It is a system for delivering power and money into fewer and fewer hands. They do not share the fundamental hope of Senator McGovern or Fred Sanderson that those will be kindly hands.

Edward C. Wolf brings this argument full circle in an essay in the 1987 *State of the World* issued by the Worldwatch Institute. He believes, as do Lappé and Collins, that traditional farmers should be helped only to enhance the ability of their native genius to grow food. He especially favors biological approaches to raising productivity that can help poor farmers to cope with the unavoidable irregularities of weather and soil type. But he says we need to preserve native farming as a resource. Native farmers have learned to use their precious soil, water, and nutrients very efficiently. Some of their techniques and wisdom, if adapted to large-scale operations, might very well protect industrial farming against the erosion and virulent pests to which it is prone. Penicillin growing in ordinary bread mould provides a model. A citation from this essay brings this final review to a fitting conclusion. It recalls not only what has been said here in praise of native agriculture, it also harks back to the examples of ingenious nomadic agriculture with which the book begins:

> The challenge for agricultural research at all levels is no longer a problem of one-way technology transfer, as so many people perceived the Green Revolution. Innovations and insights that help raise agricultural productivity will flow in both directions—between researchers and farmers, between developing and industrial countries. Success in the low-productivity fields of the third world can suggest new ways of managing agricultural resources that farmers in Iowa or France could use as well (p. 24).

CHART 5.3   Seven Fundamentals of Food Reliance

Local control

Self-improvement

Surplus-only trade

Education in nutrition

Agricultural self-reliance

Subordination of industry to agriculture

Coordinate agriculture with social reform

Conclusion

The movement of agricultural products across borders is a good topic to conclude with because almost all of the issues we have considered so far come up again. The issues become more weighty as we find the numbers of hungry people increasing as well as the tonnage of food-stuffs involved. So do the social and environmental consequences of our agricultural principles and policies. We need to double the caution with which we intervene in the ebb and flow of agricultural life.

Here especially we face the danger that short-term benefits might cause long-term harm, as we see in the dependency on American food that hinders a country like Algeria from coming to stand on her own two feet. The connections become harder to see. The intervening country will tend to have its own short-term self-interest in mind, precisely because its citizens will usually not have to live with the long-term social and environmental changes intervention might cause in a country possibly far away. Furthermore, often short-term goals will be very well intentioned. Is Senator McGovern to be faulted for his generous instincts? Does Braun's critique of the Food for Peace program in Algeria prove Theodore Schultz's ideas right? Even if they do, how do we account for the many Algerians this program might have saved from starvation?

Once again we end wondering whether there are compromises between the contending perspectives in this review. On the third kind of intervention, might it be possible to use the Food for Peace program sometimes for immediate famine relief, if we make sure any continuing program of support, be it in the form of food or expertise or materials, strives to develop both industrial and traditional agriculture in third-world countries at the same time?

Another way to ask this question is to ask whether it might be possible to develop a two-tier agriculture in the third world. Can traditional peasant agriculture, with all of its wisdom and stability, coevolve with the development of large industrialized farms growing crops for export? In China the *baogan* program described in the third case describes one possible model. A farmer receives a small personal plot to farm in traditional ways in the spare time between duties on a large, industrial, collective farm. Many reports have indicated that in virtually every instance, farmers produce more efficiently, if on a small scale, on the smaller plot. The model could be adapted for use in Kenya. The peasants on their shambas could continue to farm primarily for domestic consumption, while the unemployed urban proletariat might go to work producing specialty crops for export on large plantations. Such a system would satisfy the deep yearnings of a North American like Briggs to honor and preserve the best of both the native shamba and the big industrial farm.

In the United States there are signs that two tiers of agriculture are emerging. The American *baogan* program consists of many homestead farms, hobby farms, and second job farms thriving alongside of large agribusiness farms. The smaller-scale farms grow speciality crops or supply local markets with fresh fruits and berries. Sometimes they can afford to grow commodity crops competitively because of their light debt burdens or low production costs. They have already begun to attract the attention of pioneering researchers such as Wes Jackson. It is on their farms that his hybrids and the virtues Wendell Berry celebrates are most likely to thrive. As Frederick Buttel pointed out, American farming remains diversified because there are many farmers willing to farm for very little monetary recompense. They love the way of life. They are happy not to have to compete with the big farms of massive investments and stupendous harvests. They realize that big-time farmers have their own dreams and their own problems.

As we see in the first four chapters of this book, those American farmers who today are most in distress are the farmers who have not decided which way to go, or who have been unable to move gracefully in one direction or another. But their distress is mild compared to the anguish of peasants or traditional farmers without the luxury of choosing for themselves the future of their farms and their farming. They could probably scarcely imagine the concept of an off-farm job to support an obsession for farming. These farmers must bow to choices made elsewhere, by governments, by consumers, and by everyone in advanced countries engaged in furthering the essential human occupation of agriculture. Their hope is the hope permeating this book that the choices we make for them and for ourselves will be well-informed and humane.

Readings

Barker, Randolph. "Renewable Resource Management in Developing Country Agriculture: Discussion." *American Journal of Agricultural Economics* (December 1984), pp. 885–87.

Borlaug, Norman E. "Accelerating Agricultural Research and Production in the Third World." *Agriculture and Human Values* 3, no. 3 (Summer 1986): 5–14.

Briggs, K. "Wheat Production and Development in Kenya." *Agriculture and Forestry Bulletin* 7, no. 3/4 (September/December 1984): 27–32.

Braun, Armelle. "Rethinking Agricultural Development: Algeria's Adjustment Process." *Ceres* 3 (1983): 39–45.

Cox, George W., and Michael D. Atkins. "International Agricultural Policy." Chap. 27 in *Agricultural Ecology.* San Francisco: W. H. Freeman and Company, 1979.

Doll, Raymond J., Glenn H. Miller, Jr., and Richard D. Rees. "The Case for International Trade." Chap. 4 in *International Trade and American Agriculture.* Kansas City, Mo.: Research Department, Federal Reserve Bank, 1981.

Editors. "The Access Route for Peasants: Still Some Roadblocks." *Ceres* 96 (1986): 23–30.

Harrison, Paul. "Interview with George McRobie: 'We're not Talking about Revolution, but about Changing the Rules.'" *Ceres* 3 (1983): 33–38.

Johnston, Bruce F., and Peter Kilby. "Agriculture in a Traditional Economy"; "Agriculture and Structural Transformation." Chaps. 1 and 2 in *Agriculture and Structural Transformation.* New York: Oxford University Press, 1975.

Kristof, Nicholas D. "The Third World: Back to the Farm." *The New York Times,* 28 July 1985.

Lappé, Frances Moore, and Joseph Collins (with Cary Fowler). "What Does Food Self-Reliance Mean?" Chap. 44 in *Food First: Beyond the Myth of Scarcity.* Boston: Houghton Mifflin Company, 1975.

McGovern, George S. "Forward"; "Prologue"; "The Challenge of Hunger." Chap. 1 in *War Against Want.* New York: Walker and Company, 1964.

Norgaard, Richard B. "Traditional Agricultural Knowledge: Past Performance, Future Prospects, and Institutional Implications." *American Journal of Agricultural Economics* (December 1984), pp. 874–78.

Paarlberg, Robert. "U.S. Agriculture and Third World Development: Harmonies or Disharmonies of Interest?" in *Agriculture, Stability and Growth,* edited by Charles E. Curry and William Patrick Nichols. Port Washington, N.Y.: Associated Faculty Press, 1984.

Poleman, Thomas T. "World Food: A Perspective." In *Food: Politics, Economics, Nutrition, Research,* edited by Phillip H. Abelson. Washington, D.C.: American Association for the Advancement of Science, 1975.

Sanderson, Fred H. "The Great Food Fumble." In *Food: Politics, Economics, Nutrition, Research,* edited by Phillip H. Abelson. Washington, D.C.: American Association for the Advancement of Science, 1975.

Schultz, Theodore W. "Value of U.S. Farm Surpluses to Underdeveloped Countries." In *Foreign Agricultural Trade,* edited by Robert C. Tantz. Ames: Iowa State University Press, 1966.

Snodgrass, Milton M., and L. T. Wallace. "World Trade of Agricultural Products." *Agriculture, Economics and Resource Management.* Englewood Cliffs, N.J.: Prentice-Hall, 1975.

Wolf, Edward C. "Raising Agricultural Productivity." In *State of the World 1987,* edited by The Worldwatch Institute. New York: W. W. Norton & Company, 1987.

Index

Acid rain, 38
Africa, 100, 106, 107
Agribusiness, 23, 29, 34, 64, 77, 104; captains of, 70
Agricultural Aviation, 85
Agricultural Improvement Society, 60
Agriculture: as leverage, 5; inner-views of, 6–7; as a zero-sum game, 7, 10; over-views of, 7–8; as adversary, 8–9, 10, 48, 77; proto-agriculture, 9; sunny style of, 10–12; oily style of, 12–13; diversification in, 13; ethics for, 13; as monoculture, 13, 14, 50; principles of, 22–25; educational system of, 23, 31, 39; within nature, 29, 47–50; the debate between culture and economics in, 32, 48, 86, 88; inevitable change in, 37–38; policies of, 39–41, 47, 49, 53, 98–100, 115; most cherished myths of, 41; subsidies of, 41–47; contextual model for, 51; intermediate technology for, 51, 65, 112–13; hidden deficit economy in, 51, 109; technological progress in, 64–66, 69–71; as a moral activity, 66, 71–73, 86; history of American agriculture, 66–67; what-if vs. how-to questions in, 75–76; animal rights in, 78–84; bioengineering in, 82–84; regenerative, 86; border crossings of, 96–98; Western industrial style of, 99–100; free trade policy for, 104–9; social costs of industrial, 109; ecology in, 110–13; fundamentals of, 113–14
Agriculture and Human Values, 25, 33, 45, 87, 89, 106
Agriculture, Department of (U.S.), 50; supports agribusiness, 73–75; sponsors growth hormone research, 83; supports bioengineering, 84
Agrologist, 85
Alford, Booker T., 79
Algeria, 104, 115, 116; and Food for Peace, 102–3
American Journal of Agricultural Economics, 110
American Journal of Alternative Agriculture, 86
American Philosophical Association, 78
American Revolution, 70
Amish, 14, 30, 37, 72, 73
Animal Health Institute, 81
Animal Industry Department of Cyanamid's Agricultural Division, 82
Animal rights, 78–84; farm animals as chattel, 78; idealistic arguments in favor of, 79; right to painless slaughter, 80; Western traditions of eating meat, 80
Antibiotics, 81–82
Asia, 106
Atkins, Michael D. *See* Cox, George W.
Atlantic, 41

baogan (household responsibility system), 94–95, 116
Bangladesh, 109
Barker, Randolph, 111, 117
Berry, Wendell, 45, 73, 108, 116; *The Unsettling of America*, 26, 33; wholesome farming, 27, 40; disease of specialization, 27; desire for cultural revolution, 28; kindly use of the land, 28; dangers of bigness in farming, 29; argument based on faith, 32; need for cultural education of farmers, 40
Best, Robert S., 85, 89